Christian Barth

High Temperature Superconductor Cable Concepts for Fusion Magnets

High Temperature Superconductor Cable Concepts for Fusion Magnets

BAND 007
Karlsruher Schriftenreihe zur Supraleitung

Herausgeber

Prof. Dr.-Ing. M. Noe
Prof. Dr. rer. nat. M. Siegel

Eine Übersicht über alle bisher in dieser Schriftenreihe
erschienene Bände finden Sie am Ende des Buchs.

High Temperature Superconductor Cable Concepts for Fusion Magnets

by
Christian Barth

Dissertation, Karlsruher Institut für Technologie (KIT)
Fakultät für Elektrotechnik und Informationstechnik, 2013
Hauptreferent: Prof. Dr.-Ing. Mathias Noe
Korreferent: Prof. Dr.-Ing. John Jelonnek

Impressum

Karlsruher Institut für Technologie (KIT)
KIT Scientific Publishing
Straße am Forum 2
D-76131 Karlsruhe
www.ksp.kit.edu

KIT – Universität des Landes Baden-Württemberg und
nationales Forschungszentrum in der Helmholtz-Gemeinschaft

KIT Scientific Publishing 2013
Print on Demand

ISSN 1869-1765
ISBN 978-3-7315-0065-0

High Temperature Superconductor Cable Concepts for Fusion Magnets

Zur Erlangung des akademischen Grades eines
DOKTOR-INGENIEURS
von der Fakultät für
Elektrotechnik und Informationstechnik
des Karlsruher Instituts für Technologie
vorgelegte
DISSERTATION
von
Dipl.-Phys. Christian Barth
geb. in: Pforzheim

Tag der mündlichen Prüfung: 10.07.2013
Hauptreferent: Prof. Dr.-Ing. Mathias Noe
Korreferent: Prof. Dr.-Ing. John Jelonnek

Acknowledgments

It would not have been possible to write this dissertation without the help and support from my colleagues of the Institute for Technical Physics (ITEP) of the Karlsruhe Institute of Technology (KIT). It has been three very rewarding years in which I learned a great many things.

Above all, I would like to thank my principal supervisor, Prof. Dr.-Ing. Mathias Noe, for his guidance and support as well as for instilling in me the qualities necessary for scientific work. Special gratitude I would also like to extend to Dr. Klaus-Peter Weiss, not only academically, for his mentoring, support and assistance but also on a personal level for barbecues, hiking trips and excursions. Although a firm believer of independent work, he has always been the one to ask for help or advice in case something goes wrong.

I cherished the pleasant working environment of the "CryoMaK" group and I am very grateful to thank all its members, Dr. Klaus-Peter Weiss, Dr. Nadezda Bagrets, Dr. Michael Schwarz, Sasha Westenfelder, Elisabeth Weiss, Alexander Ehrlich, Valentin Tschan, Bianca Purr and Christoph Bayer for their help with my Ph.D. project and for conveying the team spirit. The close cooperation with the group "Superconductors - Materials and Power Applications" has been immensely helpful; I am especially thankful to Dr. Wilfried Goldacker, Dr. Sonja Schlachter, Andrej Kudymow, Dr. Michal Vojenčiak and Dr. Anna Kario for giving advice, supplying equipment and samples as well as helping with measurements.

The experiments would not have been possible without the technical assistance from Christian Lange, Frank Gröner, Markus Hollik and Mirko Gehrlein who provided components, designed circuits and were always able to repair broken equipment. Furthermore, I want like to acknowledge Dr. Walter Fietz, Dr. R. Heller, Dr. Sandra Drotziger and Pauline Leys for helping with presentations as well as for proofreading publications and the dissertation.

The contributions of Dr. Danko van der Laan and Dr. Makoto Takayasu to this work have been immensely appreciated; I especially thank them for the interesting discussions, for taking the time to visit and for the joint measurements.

During these three years, I enjoyed the scientific and philosophical discussions, swimming and bike trips and other after work activities together with Alexander Winkler, Andre Berger, Olaf Mäder, Enrico Rizzo, Phillipp Krüger, Florian Erb, Sebastian Hellmann, Oliver Näckel and Vishnuvardhan Gade. You guys have really been appreciated. Finally, I would like to thank my family and friends, for their patience, their support, as well as for the many shared activities.

Kurzfassung

In den Magnetspulen von Kernfusionsreaktoren kommen Supraleiter zum Einsatz um die Verluste zu minimieren und um die notwendigen Stromdichten zu erreichen. In bereits realisierten Fusionsexperimenten und im Internationalen Thermonuklearen Experimentellen Reaktor (ITER) werden die Tieftemperatursupraleiter Niob-Titan (NbTi), in Bereichen mit niedrigen bis mittleren magnetischen Flussdichten, und Niob-Zinn (Nb_3Sn) bei hohen Flussdichten verwendet. In den Höchstfeldbereichen der Magnetspulen zukünftiger Fusionsreaktoren bieten die Hochtemperatursupraleiter (HTS) Seltene-Erden-Barium-Kupferoxid (*RE*BCO) Bänder bedeutende Vorteile. Ziel dieser Arbeit ist die systematische Untersuchung von Hochtemperatursupraleiterkabeln mittels Simulationen und Experimenten. Die Kabel werden verglichen im Hinblick auf die Nutzung als elektrischer Leiter in Fusionsmagneten. Hierbei stehen die thermischen Eigenschaften der Materialien, die mechanischen Eigenschaften der Kabel und deren Skalierbarkeit zu den, für Fusionsmagnete notwendigen Strömen und magnetischen Flussdichten, im Vordergrund.

Die thermische Ausdehnung verschiedener Struktur-, Isolations- und Füllmaterialien wurden gemessen und mit der thermischen Ausdehnung von *RE*BCO Bändern verglichen. Niedrige Differenzen der thermischen Ausdehnungen sind notwendig um mechanische Spannungen zu vermeiden und die Supraleiterbänder vor Beschädigungen zu schützen. Dies ist für Struktur- und Kompositmaterialien, wie glasfaserverstärkter Kunststoff (GFK) gegeben. Es wurde gezeigt, dass die thermische Ausdehnung von Epoxidharz beim Mischen mit Füllern minimaler Ausdehnung (z.B. Quarzpulver) an die von *RE*BCO Bänder angeglichen werden kann.

Die thermische Leitfähigkeit der *RE*BCO Bänder und von GFK ist stark richtungsabhängig. Die thermische Leitfähigkeit in Richtung des Bandes ist drei Größenordnungen höher als senkrecht zur Bandebene. Für HTS Kabel bedeutet dies, dass die effektive Kühlkontaktfläche von der Anordnung der *RE*BCO Bänder abhängt. Der Kontakt zur Kühlung ist besser in vollständig transponierten Kabeln im Vergleich zu in Stapeln oder in Lagen angeordneten Supraleitern. Basierend auf diesen Materialuntersuchungen, wurden mehrere HTS Kabelkonzepte im Detail untersucht.

Zunächst wurde mittels der Finiten Element Methode (FEM) der Einfluss der Mäandergeometrie auf die mechanischen Eigenschaften und auf die Stromtragfähigkeit der "Roebel Assembled Coated Conductor" (RACC) Kabel untersucht. Eine optimierte Geometrie wurde vorgeschlagen, welche bei nahezu gleichem Supraleiterbedarf, die maximalen mechanischen Spannungen signifikant reduziert und Stromtragfähigkeit erhöht. RACC Kabel sind empfindlich gegenüber mechanischen Belastungen senkrecht zur Kabelrichtung und Bandebene. Ohne zusätzliche mechanische Stabilisierung können bereits niedrige senkrechte Kräfte das Kabel beschädigen. Die mechanische Festigkeit von RACC Kabeln kann durch Fixieren der *RE*BCO Bänder mittels Pressung oder Epoxidharzimprägnierung erhöht werden. Weiterhin wurde gezeigt, dass eine stufenförmige Kontaktanordnung gleichmäßige Kontaktwiderstände und homogene Stromverteilung im Kabel erzielt.

Zweitens wurde die Stromtragfähigkeit einer "Conductor on Round Core" (CORC) Kabelprobe bei verschiedenen magnetischen Hintergrundfeldern und Temperaturen gemessen. Im Eigenfeld ist die Temperaturabhängigkeit des kritischen Stromes dieser Probe vergleichbar mit der von einzelnen *RE*BCO Bändern. Alle supraleitenden Übergänge sind abrupt, mit hohen n-Werten, unabhängig von magnetischem Feld und Temperatur. Selbst bei zyklischer mechanischer Belastung wurde keine Degradation der CORC Kabelprobe festgestellt.

Drittens wurde eine "Twisted Stacked-Tape Cable" (TSTC) Probe feld- und temperaturabhängig charakterisiert. Es wurde eine permanente Reduzierung der Stromtragfähigkeit der Probe während des ersten Anstieges des magnetischen Hintergrundfelds festgestellt. Die Degradation sättigte und veränderte sich in darauffolgenden Feldzyklen nicht.

Fünf HTS Kabelkonzepte wurden verglichen und hinsichtlich ihrer Eignung als elektrische Leiter in Fusionsmagneten bewertet. Alle untersuchten HTS Kabelkonzepte, mit Ausnahme von RACC Kabeln, können die elektrischen und magnetischen Randbedingungen von Fusionsmagneten erfüllen. Von den untersuchten Proben zeigte das CORC Kabel die besten mechanischen und elektrischen Eigenschaften. Jedoch ist Aufgrund der Anordnung der *RE*BCO Bänder in Schichten, eine Transposition der Supraleiter in CORC Kabeln nur partiell gegeben.

Abschließend wurden zwei HTS Wicklungspakete für Toroidalfeldspulen berechnet, eines entsprechend den Randbedingungen von ITER, das andere für das Fusionskraftwerkkonzept DEMO1 (2012). Bei gleichem Sicherheitsfaktor sowie gleicher Menge an Strukturmaterialien, elektrischer Stabilisierung und Kühlquerschnittfläche kann die gesamte Querschnittfläche des ITER Wicklungspakets um 8,9 % reduziert werden, die des Wicklungspakets des Fusionskraftwerks DEMO1 sogar um 12,8 %.

Contents

7 Summary

A Annex

B Designations and abbreviations

C Index of symbols

D Bibliography

1 Introduction and motivation

Since Einstein's discovery of the mass-energy equivalence ($E = m \cdot c^2$) in 1905 [Ein05b, Ein05a], nuclear reactions are suspected to be powerful energy sources. In 1938, nuclear fusion was discovered as the energy source of the stars by Hans Bethe [Bet39]. Since then, mankind dreamed of harnessing that power for electrical energy generation. Fusion fuels, such as deuterium and tritium, which can be bred from lithium, are abundant on earth [HB01]. Additionally, controlled fusion reactions are inherently safe and produce neither greenhouse gases nor radioactive waste which has to be stored for millenia [Oak]. Thus, nuclear fusion is seen as a sustainable energy source of the future.

In fusion reactions, close proximity of the reactants is necessary. The repellent Coulomb potential has to be overcome, requiring temperatures in the 100 million degree range [Cul12]. At such high temperatures, matter is in its fourth state, the plasma. Nuclei and electrons are disassociated. The "positively charged nuclei swim in a 'sea' of freely-moving negatively charged electrons"[a]. To use fusion as an energy source, these conditions have to be controlled and maintained over long time periods. Contact-less confinement of the plasma is necessary, keeping the fusion reaction alive and minimizing the thermal energy loss to surrounding areas. As plasma consists of charged particles, it can be influenced by magnetic fields. Magnetic confinement using helical toroidal magnetic fields has been shown to be possible. The most promising are the Tokamak and Stellarator coil configurations. For significant fusion powers, high magnetic fields in large volumes are required. These necessary magnetic fields cannot be generated efficiently with conventional, resistive, magnets.

A solution is offered by superconductivity, which was discovered by Heike Kamerlingh Onnes in 1911 [Onn12]. In superconductors, direct currents (DC), up to the critical current of the material, can be transported without any losses. This drastically increases the efficiency and, due to much higher current densities, leads to significantly more powerful magnets. In fusion, superconducting magnets are used starting with the "Baseball I" in 1965 [HLCB82]. Since then, the superconducting fusion experiments have grown larger and larger [Vel08], nearing the point of an energy gain of the nuclear fusion reaction. This will be achieved with the International Experimental Thermonuclear Reactor (ITER), at present the largest fusion experiment [ITE09a]. ITER utilizes the low temperature superconductor (LTS) materials niobium-titanium (NbTi)

---

[a]from Wikipedia, the free encyclopedia: State of matter

and niobium-tin (Nb_3Sn) to generate the necessary magnetic fields [MBB09]. The magnets of ITER are cooled with liquid helium (He) preserving the superconducting properties. Low temperature superconductors are an established technology and are not only used in fusion, but also in magnetic resonance imaging (MRI) devices and accelerator magnets.

In the next step of controlled fusion research, ignition of the fusion reaction is necessary to further increase the energy output. A demonstration reactor (DEMO) is planned, to show the applicability of nuclear fusion for electrical power generation [MCS05]. For DEMO, an increase in the magnetic fields and of the plasma volume is necessary, thus requiring even more powerful magnets [Zoh10]. As the present conductor technology is already at its limit with the ITER magnets, advanced solutions are needed for DEMO.

Second generation high temperature superconductors (HTS), also referred to as coated conductors or rare-earth-barium-copper-oxide ($Rare\text{-}Earth_1\,Ba_2Cu_3O_{7\text{-}x}$ or $REBCO$) tapes, are promising candidates in future fusion magnets. For example, these materials possess vastly superior mechanical properties and current densities at very high magnetic background fields compared with LTS [Haz10, Bru12, Supd, Sel11b]. Additionally, high temperature superconductors can be operated at higher temperatures. This enables simplifications in the cryostat design, in electrical cold testing, and in the cooling of the magnets. Furthermore, helium may be avoided as cooling agent in HTS magnets as it is scarce on earth. However, HTS magnets are a new technology. Up to now, there are no fusion relevant HTS magnets available. In contrast to the round wires of low temperature superconductors, $REBCO$ tapes are commonly of flat geometry requiring new concepts for cabling.

There are several concepts of high current $REBCO$ cables to be used in fusion magnets. Major concepts are the Roebel Assembled Coated Conductor (RACC) cable [GNK06], the Coated Conductor Rutherford Cable (CCRC) [SDR10], the Conductor on Round-Core (CORC) cable [Laa09], the Twisted Stacked-Tape Cable (TSTC) [TCBM11, TMB] and a cable concept consisting of Round Strands Composed of Coated Conductor Tapes (RSCCCT) [UWB13b]. These HTS cable concepts are developed by various research institutions and companies around the world [MBT11].

The main objective of this work is an systematical investigation and optimization of HTS cables for fusion magnets. Starting with the constituent materials, the cables are analyzed through experiments and simulations. The Roebel Assembled Coated Conductor (RACC) cable concept is optimized, improving the mechanical properties, the current carrying capabilities and the homogeneity of the current distribution. The performance of RACC cables and other HTS cables is investigated at different magnetic background fields and temperatures, through measurements in the FBI (force F - field B - current I) test facility. All HTS cable concepts are assessed regarding their applicability as conductors in fusion magnets at the boundary conditions of the ITER toroidal field (TF) coils. The investigations are focused on the mechanical properties, on methods to contact the *RE*BCO tapes and on the scaling of the cables to the electrical currents necessary for fusion magnets (68 kA at 12 T background fields). Two HTS winding packs, one for ITER and one for a fusion power plant are extrapolated using CORC cables as conductors. The cross sectional areas are compared with the corresponding LTS winding packs highlighting the applicability and benefits of HTS as conductor in future fusion magnets.

In **chapter 2**, superconductivity is introduced and the basic phenomena are described. Characteristic values of superconductors and principles regarding operation and stability in technical applications are given. The composition, manufacturing processes, properties and application areas of the commercially available technical superconductors are described and compared. The comparison mainly focuses on the aspects essential in fusion magnets, the critical current density in high magnetic background fields and the mechanical properties.

Superconductors enable large and efficient high field magnets, necessary for magnetically confined nuclear fusion. Superconducting fusion magnets are described in detail in **chapter 3**, starting with the basics of nuclear fusion, leading to the two major magnetic coil arrangements, the Tokamak and the Stellarator. The history of superconducting fusion magnets is shown from the "Baseball I" in 1965 to ITER, today's largest fusion experiment. State-of-the-art superconducting fusion magnet technology is illustrated in the example of ITER. Going beyond ITER, the main parameters of fusion power plants are specified and their impact on the superconducting coils are evaluated. Challenges of future fusion power plants are considered and balanced with the potential of improvements using high temperature superconductors.

To use high temperature superconductors in fusion magnets, cables are necessary. HTS cables are new technologies due to the significant differences in shape and properties of LTS and HTS materials. These cables have to be analyzed in detail, ranging from their constituent materials, over the arrangement of the HTS tapes and their performance in high magnetic background fields to the scalability to fusion magnet's currents.

For the materials of HTS cables, the thermal properties are highly important due to the huge temperature differences between room temperature (RT) and operating temperatures, which may be as low as 4 K. In **chapter 4**, the thermal expansion and the thermal conductivity of the constituent materials of HTS cables, superconductors, structural materials, electrical insulation and solders, glues and resins are investigated. The thermal properties of these different material types diverge in orders of magnitudes,. All materials are assessed regarding two main design rules. Firstly, mismatches of the thermal expansions have to be minimized in stress sensitive components to avoid damage during temperature changes. Secondly, the thermal conductivities in materials separating the superconductors and the cooling agents have to be as high as possible, maximizing the thermal transport and improving the effective cooling contact area. Based on these investigations, structural materials, insulating materials, and resins are recommended for the use in HTS cables.

Designing superconductor cables is more than just choosing suitable materials, the arrangement of the superconductors and the connections to the current leads are significant factors, too. In **chapter 5**, one HTS cable concept, the Roebel Assembled Coated Conductor (RACC) cable is analyzed and optimized. RACC cables consist of meander shaped *REBCO* tapes which are combined into a cable using the Roebel assembling method [Sta12]. The meander shape, which is the distinct geometry of this cable type, is identified in a first step as strongly influencing the performance of the cable. The geometry is parametrized and the impact of the parameters on the mechanical stability and the current carrying capabilities is investigated systematically leading to an improved meander shape. In a second step, homogeneous contact resistances between all tapes and the current leads are demonstrated as being vital in achieving a uniform current distribution in DC applications. Contact area, soldering method and pressure during soldering are identified as the main governing parameters. A new contact type is suggested and compared with the state-of-the-art contacts of RACC cables.

These finding are not just limited to the RACC cables. In all HTS cable concepts, the mechanical properties, the current carrying capabilities in magnetic fields, contacts and scalability have to be considered at the boundary conditions of fusion magnets. In **chapter 6**, the major HTS cable concepts, Roebel Assembled Coated Conductor (RACC) cables, Coated Conductor Rutherford Cables (CCRC), Conductor on Round Core (CORC) cables, Twisted Stacked-Tape Cables (TSTC) and cables consisting of Round Strands Composed of Coated Conductor Tapes are described and compared. The FBI (force F - field B - current I) test facility is extended with a temperature variable insert, allowing the characterization of HTS cable samples from 4.2 K to 77 K in magnetic background fields up to 12 T. The scaling of the cables to fusion magnet relevant currents is investigated using the conditions of the ITER toroidal field (TF) coils as reference. To demonstrate the applicability and benefits of HTS cables as conductors in fusion magnets two winding packs are extrapolated and compared with LTS. One for the inner leg of the ITER toroidal field coils and the other for a fusion power plant. The safety factor, the void areas, the amount of electrical and mechanical stabilization are kept constant in these calculations.

This work is summarized and concluded in **chapter 7**.

2 Superconductors

2.1 Superconductivity

Superconductivity was discovered by Heike Kamerlingh Onnes in 1911 by measuring the specific resistivity of mercury cooled with liquid helium [Onn12]. He found a vanishing electrical resistivity of mercury at a certain temperature. Below this temperature the material behaved as a "superior conductor" and he called this effect superconductivity.

2.1.1 Type I and type II superconductors

In the superconducting state there is no electrical resistivity for direct current (DC) applications. Additionally, magnetic fields are expelled from the inside of the superconductor in the superconducting state. The material behaves as a perfect diamagnet except in a very thin region near the border where the magnetic flux can penetrate. This is called the Meissner phase [MO12].

There are several elements which exhibit phase transitions to such superconducting states at certain temperatures. Superconductivity exists in the following elements at ambient pressure: lead (7.196 K), lanthanum (4.88 K), tantalum (4.47 K), mercury (4.15 K), tin (3.72 K), indium (3.41 K), palladium (3.3 K), chromium (3.0 K), thallium (2.38 K), rhenium (1.697 K), protactinium (1.4 K), thorium (1.38 K), aluminum (1.175 K), gallium (1.083 K) and several other with even lower transition temperatures [SUPb]. In these superconductors even very small currents or magnetic fields cause a breakdown of the superconducting state and a sharp transition to normal conduction. The superconducting elements are the type I superconductors and are unsuitable for technical power applications [Kom95, p. 30]. The transport currents and magnetic fields, while remaining in superconducting state, are just too small.

In 1933 a different type of superconductivity, the type II superconductivity, was discovered in alloys and compounds [SUPc]. In type II superconductors, low magnetic fields are completely expelled and there is also a superconducting Meissner phase. Increasing the magnetic field causes a breakdown of the Meissner phase. In contrast to the type I superconductors, this is not a transition to normal conduction, it is a transition to a second superconducting phase instead [Abr57]. The transition takes place at a lower critical magnetic field B_{c1}. In this second superconducting phase, the Shubnikov phase, magnetic flux penetrates the material. In contrast to normal conduction, the flux is not distributed homogeneously, but penetrates the material in quantized flux tubes. Each magnetic flux tube is stabilized by a superconducting circulating current. Increasing the magnetic field also increases the number of flux tubes. This is still a superconducting phase, there is no electrical resistivity. If a current I is transported through at type II superconductor in the Shubnikov phase, Lorentz forces act on the flux tubes as shown in **equation 2.1** [Kom95, p. 31 ff.].

$$F_{\mathrm{L}} = B_{\mathrm{i}} \cdot I \cdot \ell. \tag{2.1}$$

The Lorentz force, F_{L}, depends on the magnetic field B_{i} in the superconductor, the current I and the length of the superconductor l. The Lorentz force, F_{L}, moves the flux tubes. This is called "flux creeping" and is shown in **figure 2.1**. Through this movement of the flux tubes and the disbanding and generation of new flux tubes at the border of the superconductor, energy is dissipated. Through flux creeping an electrical field occurs in a superconductor in the Shubnikov phase, the "flux creeping resistance" [Kom95, p. 31].

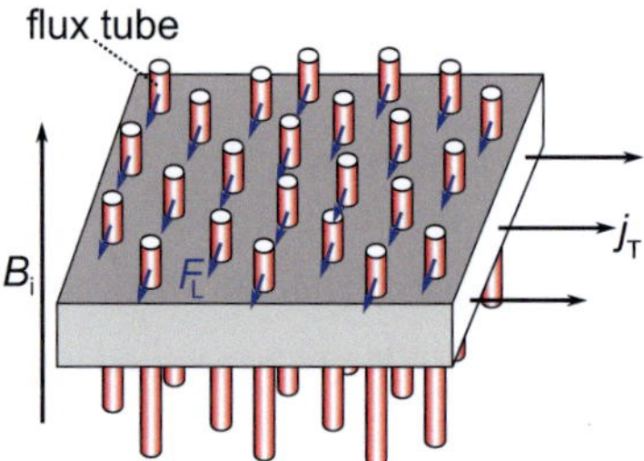

Figure 2.1: Flux creeping in type II superconductors. A transport current density j_{T} results in Lorentz forces F_{L} at the flux tubes in an internal magnetic field B_{i}. Picture after [Kom95, p. 30].

Flux tubes are normal conducting on the inside. Therefore it is energetically favorable if a flux tube is running through a defect in a superconductor as such a defect is also normal conducting. The flux tube is pinned to the defect. Mechanical work is needed to displace the flux tube a distance x away from the defect as shown on the left of **figure 2.2**. This means there is a certain force which pins the flux tube to a defect of a thickness t. This is the pinning force F_{p} [Kom95, p. 32], which can be calculated according to **equation 2.2**.

$$F_{\mathrm{p}} = \frac{e \cdot t}{x} \tag{2.2}$$

The pinning force F_{p} is always opposed (anti parallel) to the Lorentz force F_{L} at each pinned flux tube. As long as the pinning force is equal to the Lorentz force the flux tube remains pinned and does not move (**figure 2.2** on the right). If the Lorentz force is increased further, flux creeping takes place and energy is dissipated in the superconductor. Thus transporting current without an electrical resistivity is only possible in the Meissner phase or the Shubnikov phase

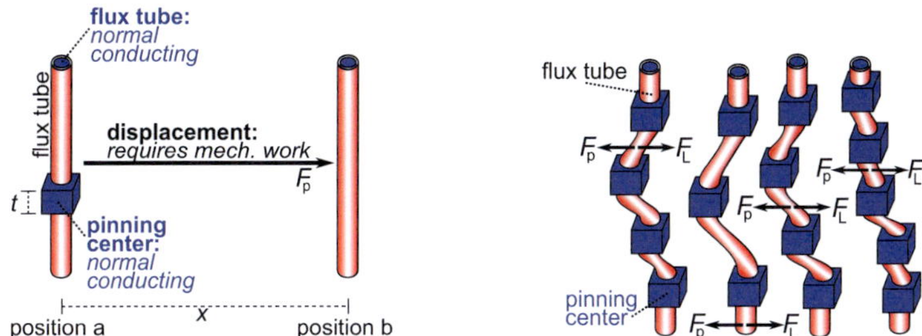

Figure 2.2: Flux pinning in type II superconductors. To displace a flux tube from a pinning center with a thickness t, a force F_p is required. This is the pinning force (left). Therefore flux tubes attach themselves to pinning centers. The resulting pinning forces F_p counter the Lorentz forces F_L (right). Pictures after [Kom95, p. 31 and p. 33].

if the pinning forces are high enough. The pinning forces can be increased by increasing the number of defects. This is possible by doping [ZGRS07] or by nano structuring [Mos00] the superconductor material.

At significantly higher magnetic fields a second phase transition takes place. At the upper critical magnetic field B_{c2} the Shubnikov phase breaks down and the alloy is in normal conducting state. The transition from the Meissner phase to the Shubnikov takes place at very low magnetic fields of only a few milli tesla (B_{c1}). The second transition occurs at much higher fields. B_{c2} is of the order of several tesla. For technical power applications, e.g. magnets, transformers, fault current limiters, rotating machines, power cables, only type II superconductors are relevant due to B_{c2} being far greater than B_{c1}. In **figure 2.3** the correlation of the magnetic flux inside materials and applied background fields is shown for normal conductors, type I superconductors and type II superconductors.

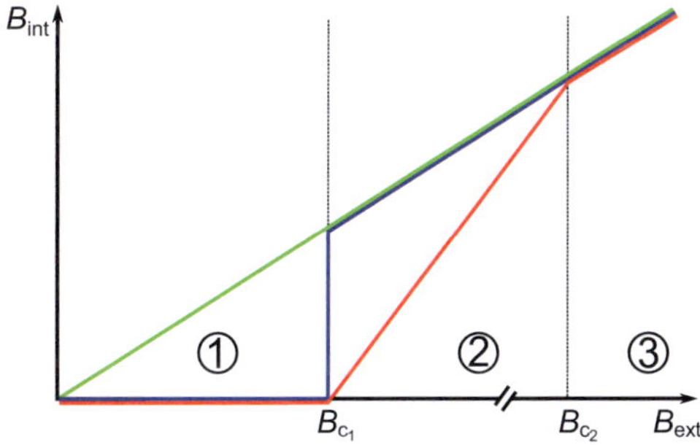

Figure 2.3: Correlation of magnetic flux density inside materials (y-axis: B_{int}) depending on magnetic background fields (x-axis: B_{ext}) in case of a normal conductor (in green), a type I superconductor (in blue) and a type II superconductor (in red). Meissner phase ①, Shubnikov phase ② and normal conducting phase ③. Picture after [Kom95, p. 24].

2.1.2 Critical values

The conditions in which a superconductor retains its superconducting properties is defined by its critical values. These critical values are the critical temperature T_c, the critical magnetic field B_c and the critical current density j_c and depend on the superconducting material. The critical values of superconductors are not constant, they are dependent on each other. The critical current for example is lower in a higher magnetic background field or at increased temperatures.

2.1.2.1 Critical temperature

The transition from the superconducting state (Meissner or Shubnikov phase) to normal conduction is not spontaneous at the critical temperature. If examined in detail, there is a transition zone around the critical temperature. To define the critical temperature of a superconductor as a single value, three different methods can be used [Gau01, Cav98] as shown schematically in **figure 2.4**.

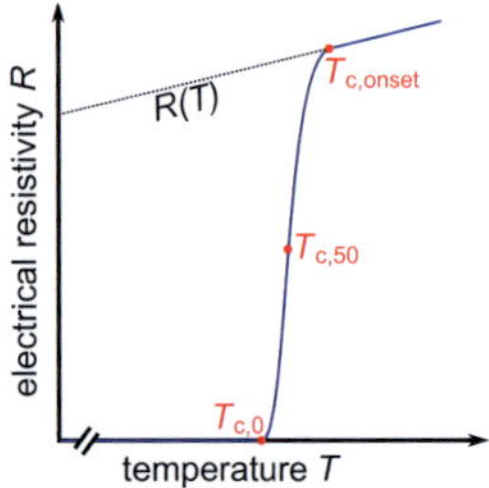

Figure 2.4: Schematic drawing of the transition of a superconductor from superconducting state (Meissner or Shubnikov phase) to normal conduction at the critical temperature. Different definitions of critical temperature $T_{c,onset}$, $T_{c,50}$ and $T_{c,0}$. Picture after [Cav98].

The critical temperature T_c is the temperature at which:

- there is a deviation from the characteristic curve of the temperature dependence of the electrical resistivity for normal conduction. This temperature is also called $T_{c,onset}$.

- the resistivity is half of the resistivity gained from an extrapolation of the characteristic curve of the temperature dependence of the electrical resistivity in normal conduction. This temperature is also called $T_{c,50}$.

- the resistivity starts to be measurable. This temperature is also called $T_{c,0}$.

2.1.2.2 Critical magnetic field

In a superconductor transition to normal conduction occurs if the magnetic field B is above a critical value. This value is the critical magnetic field B_{c2} (or B_{c1} in type I superconductors), in technical superconductors commonly referred to as B_c. It depends on the superconductor material and on the temperature. The temperature dependence of B_c can be approximated using **equation 2.3** [Kom95, p. 35][Dre95, p. 3].

$$B_c(T) = B_c(0\,\mathrm{K}) \cdot \left(1 - \left(\frac{T}{T_c}\right)^2\right) \qquad [2.3]$$

The field can either be applied from the outside (external magnetic field or background field) or can be generated by a current through the superconductor (self field). This means a current in the superconductor reduces the background field B_{ext} at which the transition to normal conduction occurs. Thus, the external magnetic field B_{ext} and the critical current density j_c are dependent on each other. Through empirical observation it has been shown that this dependency can be approximated at constant temperature using **equation 2.4** [KHS63].

$$j_c = \frac{\alpha_B}{B_{ext} + B_0} \qquad [2.4]$$

In this equation, α_B and B_0 are both material dependent constants. The material constant B_0 is of a magnitude of a few milli-tesla. In magnetic background fields B_{ext} of several tesla, the material constant B_0 is negligible. Thus in such high fields, the critical current density j_c is inversely proportional to an applied magnetic background field B_{ext} as shown in **equation 2.5** [Dre95, p. 8].

$$j_c \propto B_{ext}^{-1} \qquad \forall B_{ext} \gg \mathrm{mT} \qquad [2.5]$$

2.1.2.3 Critical current density

In the superconducting state the electrical resistivity vanishes. Because of this, there is no electric field along the superconductor. At the transition to normal conduction an electrical field E appears. The electrical field appears gradually at the transition. Through empirical observation it has been shown that the characteristic $E - j$ curves of superconductors can be described near the transition to normal conduction with a power function. The corresponding equation is shown in **equation 2.6** and is commonly referred to as the "Power Law" of superconductors [Buc94, See98, Lee01, Wal74, Kom95].

$$E = E_c \cdot \left(\frac{j}{j_c}\right)^n \qquad [2.6]$$

In an ideal superconductor the transition from superconducting state to normal conduction would be instantaneous. Ideal superconductors can be described with the Bean model of the critical state [Bea62]. With the Power Law these models can be generalized to allow gradual transitions essential for describing technical superconductors. A characteristic $E - j$ curve of a technical superconductor is schematically shown on the left of **figure 2.5**. Measuring the

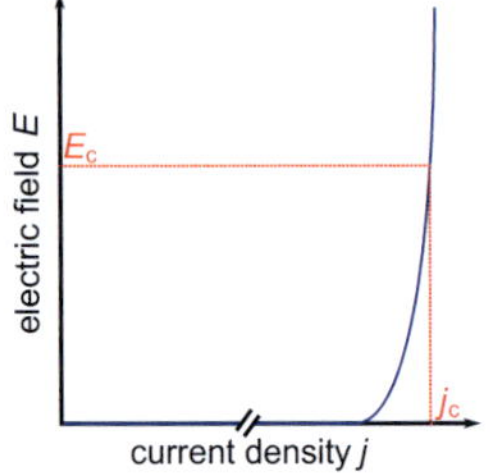
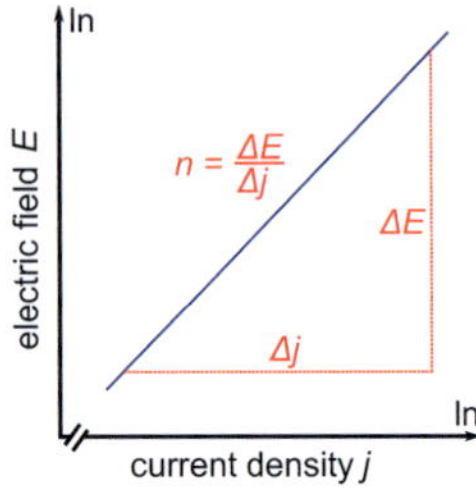

Figure 2.5: Schematic drawing of electrical field E (y-axis) and current density j (x-axis) dependency in superconductors. As shown on the left the critical current density j_c of superconductors is defined as the current density j at which a critical magnetic electric field E_c occurs at the superconductor. In a double logarithmic representation the slope corresponds to the n-value n of a superconductor (on the right). Pictures after [Kom95, p. 84].

electrical field is a common method to determine the critical current density j_c. The critical current density j_c is defined as the current density j, which generates the critical electric field E_c along a superconductor. For the low temperature superconductors, niobium-titanium (NbTi) and niobium-tin (Nb$_3$Sn), the critical electric field E_c is set to $0.1\,\mu\text{V}\,\text{cm}^{-1}$ by international standards [DKE07a, DKE07b], for the first generation high temperature superconductors bismuth-strontium-calcium-copper-oxide (Bi$_2$Sr$_2$Ca$_2$Cu$_3$O$_y$ or BSCCO 2223) tapes E_c is set to $1\,\mu\text{V}\,\text{cm}^{-1}$ instead [DKE06]. The critical electric field E_c of the second generation high temperature superconductors, rare-earth-barium-copper-oxide (Rare-Earth$_1$Ba$_2$Cu$_3$O$_x$ or REBCO) tapes, has not yet been standardized [WBBG11, BWGS11]. However, a critical electric field E_c of $1\,\mu\text{V}\,\text{cm}^{-1}$ is commonly used with these materials [LLG11, TMB11, GFK09].

The steepness of the transition from superconducting state to normal conduction at the critical electric field is described by the exponent or power factor n of the Power Law (**equation 2.6**). It is called the "n-value" of a superconductor [EL93] and can be obtained as the slope near j_c in a double logarithmic display [GFS03, TH05] as shown on the right of **figure 2.5**. The slope in the region of the critical electric field E_c to $10 \cdot E_c$ is used commonly to determine the n-value [TH05]. A high n-value results in a sharp transition from superconducting state to normal conduction [NJW99].

2.1.2.4 Operating parameters

If the critical values, the critical temperature T_c, the critical magnetic field B_c and the critical current density j_c are plotted on the axes of a coordinate system, all critical values form a surface as shown on the left of **figure 2.6**. For each point at or below the surface the superconductor is in superconducting state. As the critical values depend on each other, an operating point can be described as $j_c(B,T)$. Above the surface the superconductor is in the normal conducting state. For stable operation, operating points have to be chosen at a certain distance below the surface to allow fluctuations of the operating current density j_{op}, the operating temperature T_{op} or the magnetic field B_{op} without leaving the superconducting state. A safe operating point therefore contains a current reserve Δj at constant magnetic field and temperature or a temperature reserve ΔT at constant current and magnetic field. This is shown schematically on the right of **figure 2.6**.

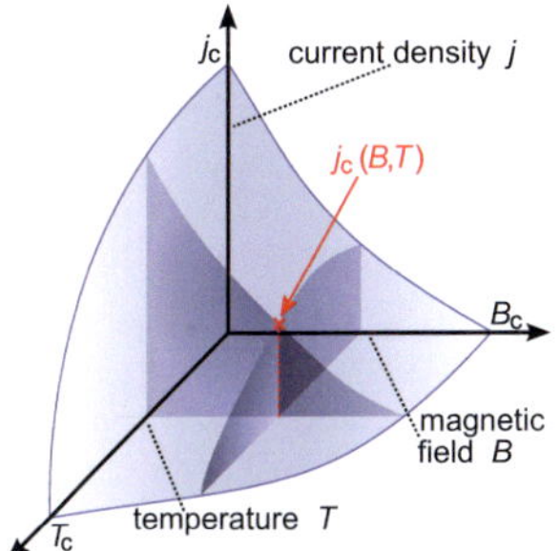

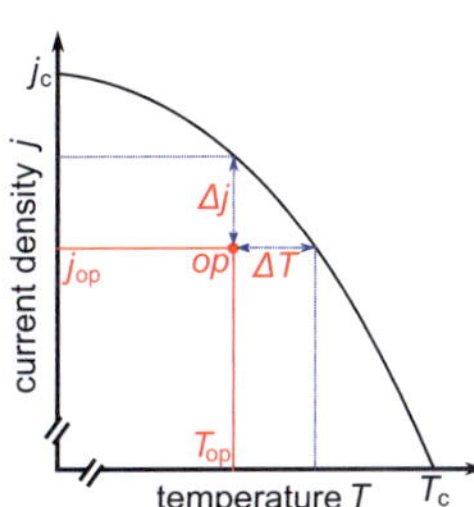

Figure 2.6: Critical values of superconductors (left): Temperatures T are displayed on the x-axis, magnetic background fields B on the y-axis and current densities j on the z-axis. Picture after [Kom95, p. 34]. Operating point with safety margins (right): The operating point (op) contains a current reserve Δj and a temperature reserve ΔT. Picture after [Iwa09, p. 357][Mae12, p. 12].

Superconductors in technical applications are equipped with electrical stabilization to carry the current if the superconductivity breaks down in the case of a quench. The operating parameters of technical superconductors are commonly described by the engineering current density j_e and the current sharing temperature T_{cs}. At constant magnetic field and temperature, the engineering current density j_e is the quotient of the critical current of the superconductor I_c and the total cross section area A_{tot} including the stabilization. The current sharing temperature T_{cs} is the temperature at which the critical electric field E_c of the superconductor (usually $0.1\,\mu V\,cm^{-1}$ for NbTi and Nb$_3$Sn or $1\,\mu V\,cm^{-1}$ for BSCCO and *REBCO*) is generated at a given current and temperature [HCD03]. At the current sharing temperature, T_{cs}, the superconductor is in the transition from superconducting state to normal conduction. The current is not transported solely by the superconductor, it is shared between the superconducting parts and the normal conducting parts (e.g. matrix, electrical stabilization or substrate). The current sharing temperature T_{cs} is similar to the measurable resistivity definition the critical temperature $T_{c,0}$, whereas "measurable" is defined by the critical electric field E_c of the superconductor [NIS03].

2.1.3 Technical superconductors

The critical values and their correlation depend on the superconducting material. For technical power applications only type II superconductors are relevant. The first superconductors used in applications were the low temperature superconductors (LTS) niobium-tin (Nb_3Sn) [MGGC54] and niobium-titanium (NbTi) [EL63, BH63, Ber]. The superconductivity of these superconductors is explained in the BCS theory proposed by Bardeen, Cooper and Schrieffer in 1957 [BCS57]. In 1986 superconductivity was discovered in lanthanum-barium-copper-oxide ($La_{2-x}Ba_xCu_1O_4$), a cuprate-perovskite ceramic material with a critical temperature of 30 K [BM86]. The critical temperature of the materials is above the upper limit of 20 K of the BCS theory. Therefore these new types of superconducting materials are often referred to as "non-BCS superconductors" or "unconventional superconductors", or due to their high critical temperatures as "high temperature superconductors" (HTS). At present there is no clear definition of which critical temperature is needed for a material to be a high temperature superconductor. The historic value of 30 K or the boiling point of liquid nitrogen (77 K) are commonly used. In this work only materials with a critical temperature above the boiling point of liquid nitrogen (77 K) are referred to as high temperature superconductors. There are 6 commonly used superconductors, 2 low temperature superconductors (LTS), 3 high temperature superconductors (HTS) and MgB_2. These are the so called technical superconductors:

- Niobium-titanium (NbTi)

- Niobium-tin (Nb_3Sn)

- Bismuth-strontium-calcium-copper-oxide ($Bi_2Sr_2Ca_1Cu_2O_y$ or BSCCO 2212)

- Bismuth-strontium-calcium-copper-oxide ($Bi_2Sr_2Ca_2Cu_3O_y$ or BSCCO 2223) as tape and bulk material

- Rare-earth-barium-copper-oxide (*Rare-Earth*$_1Ba_2Cu_3O_{7-x}$ or *RE*BCO) as tape and bulk material

- Magnesium-diboride (MgB_2)

There are significant differences in the critical values, critical temperature T_c, critical magnetic field B_c, critical current density j_c, and in the dependence of these values on each other. The critical temperature T_c and the upper critical magnetic field B_{c2} is shown for the technical superconductors in **table 2.1**. In the metallic superconductors (NbTi, Nb_3Sn and MgB_2), the critical temperature T_c and the upper critical magnetic field B_{c2} are influenced by the composition of the alloy [Wil83, p. 287].

The engineering current density j_e, as a function of applied magnetic field B is shown in **figure 2.7** for a constant temperature of 4.2 K. The engineering current density determines the

Table 2.1: Critical temperature T_c, upper critical magnetic field B_{c2} and application area of technical superconductors. Data from: NbTi [Kom95, p. 70][Wil83, p. 287]; Nb$_3$Sn [Kom95, p. 70][VKS07]; BSCCO 2212: [IM91, WTM10]; BSCCO 2223 [LEH01, MYI07, Bal02]; *RE*BCO [WAT87, SMM07, WTM10] and MgB$_2$ [NNM01, Eis07, PTM04].

material	$T_c(0)$ / K	$B_{c2}(0)$ / T	application area
NbTi	9.6	12 - 15	medium fields: magnets up to 9 T
Nb$_3$Sn	18	25 - 29	high fields: magnets up to 15 - 20 T
BSCCO 2212	95	175 - 225	highest fields: inserts for magnets above 20 T
BSCCO 2223	107	107	medium fields: e.g. power cables and current leads
*RE*BCO	92 - 95	120 - 250	highest fields: inserts for magnets above 20 T
MgB$_2$	35 - 39	14 - 40	low to medium fields: e.g. wiring in satellites

applicability of the superconductor materials. At 4.2 K, niobium-titanium (NbTi) is commonly used up to 9 T. When cooled with superfluid helium (1.8 K), up to 11 T are possible [Kom95, p. 69]. At higher fields, niobium-tin (Nb$_3$Sn) is the reference material, useable up to 15 - 20 T at 4.2 K [Kom95, p. 69]. Fields above 20 T are possible with BSCCO 2212 round wires and rare-earth-barium-copper-oxide (*RE*BCO) tapes. The field dependence of engineering current density of these superconductor materials is low in high fields, revealing an applicability at even above 30 T [WTM10].

BSCCO 2223 tapes are commonly used in low to medium magnetic field applications as HTS power cables [MYI07] or current leads [Bal02]. Magnesium-diboride (MgB_2) can be used in areas with low to medium magnetic fields e.g. the wiring of satellites [PTM04], it is only shown for completeness and is not discussed in the following.

2.2 Low temperature superconductors

At present, most superconducting applications utilize the low temperature superconductors niobium-titanium (NbTi) and niobium-tin (Nb$_3$Sn).

2.2.1 Niobium-titanium (NbTi)

Niobium-titanium (NbTi) is an alloy of niobium and titanium. It was discovered as a type II superconductor in 1962 [EL63]. NbTi superconductors are manufactured from pure melts of niobium and titanium. The melt is cast to rods which are inserted in copper tubes. These are repeatedly heated and extruded, stacked into copper tubes and twisted. This process is repeated several times. A final cold drawing and swaging forms round superconducting wires consisting of μm thick niobium-titanium filaments which are embedded in a copper matrix [Kom95, p. 72].

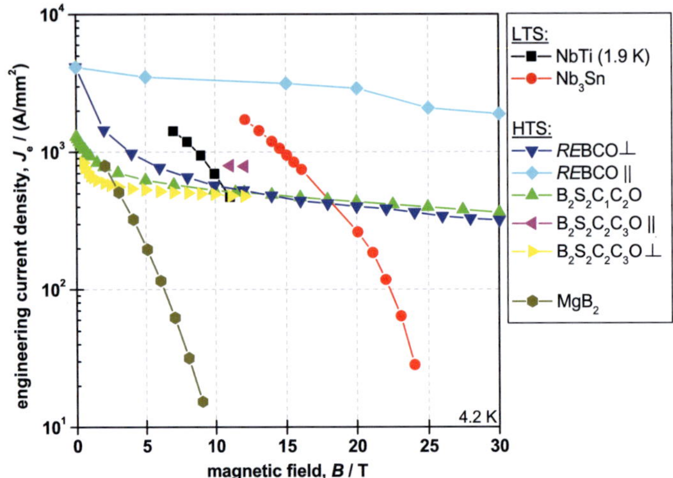

Figure 2.7: Engineering critical current density (J_e) vs. applied magnetic field (B) of technical superconductors at 4.2 K. Data from [Lee11].

The cross section of a NbTi superconductor is shown schematically in **figure 2.8**.

Depending on the desired shape of this multifilament wire, different niobium-tin to copper ratios are used. NbTi is a ductile material with very good mechanical properties, far superior to Nb_3Sn. The influence of mechanical strain on the current carrying capabilities is low, the critical strain limit in the percent range. The handling and manufacturing of NbTi is much easier than that of the other low temperature superconductors as NbTi is insensitive to bending and can be directly wound to coils ("react & wind" method). NbTi is utilized in applications with intermediate magnetic fields (9 T at 4.2 K) e.g magneto-resonance-imaging (MRI) devices, accelerator magnets or some fusion magnets, e.g. the coils of Wendelstein 7-X (W7-X) [Sap00], the coils of the Japanese Tokamak 60 Super Advanced (JT-60 SA) [IBK10, YTK10] and the poloidal field coils of ITER [LSI11].

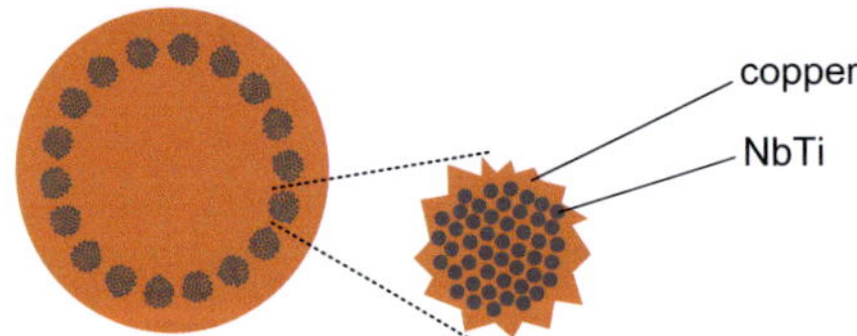

Figure 2.8: Schematic cross section of niobium-titanium (NbTi) superconductor. Picture after [Kom95, p. 74][Wil83, p. 291].

2.2.2 Niobium-tin (Nb$_3$Sn)

Niobium-tin is an inter-metallic compound of the A15 phase composition [SA90]. Nb$_3$Sn was discovered as a superconductor in 1954 [MGGC54]. It can be made using different processes e.g. internal tin process, powder-in-tube process or bronze route. The process determines not only the maximal achievable critical current density and the in-field performance, but also the mechanical properties. To achieve maximal current density an internal tin process is commonly used, although niobium-tin manufactured with a bronze-route usually has superior mechanical properties. In all manufacturing processes a final heat treatment is required to form the A15 phases. Nb$_3$Sn from these two manufacturing processes is shown in a schematic cross section view in **figure 2.9**.

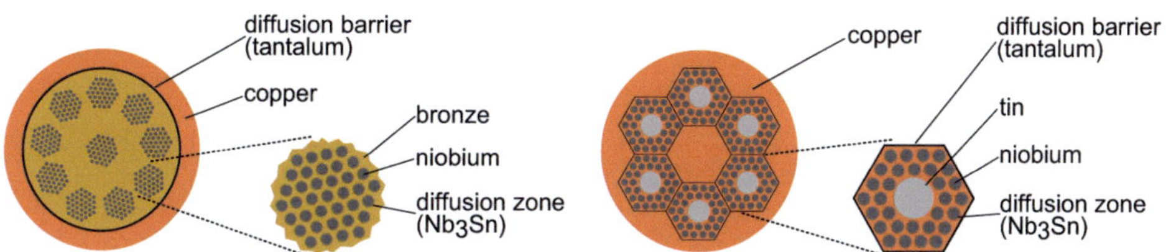

Figure 2.9: Schematic cross section of niobium-tin (Nb$_3$Sn) superconductor manufactured with the bronze route (left) and with the internal tin process (right). Pictures after [Kom95, p. 78][Wil83, p. 299].

Niobium-tin is a low mechanical strength material with low critical stress and strain limits as shown on the left of **figure 2.10**. The critical strain limit is usually less than 0.4 % unless pre-strain is applied during manufacturing. The critical stress limit is in the 100 - 200 MPa range, depending on the manufacturing process and the composition. After the superconducting A15 phases are formed, Nb$_3$Sn is, as are most A15 superconductors, very brittle. Even slight bending can cause irreversible damage to the reacted Nb$_3$Sn strands, making it very challenging

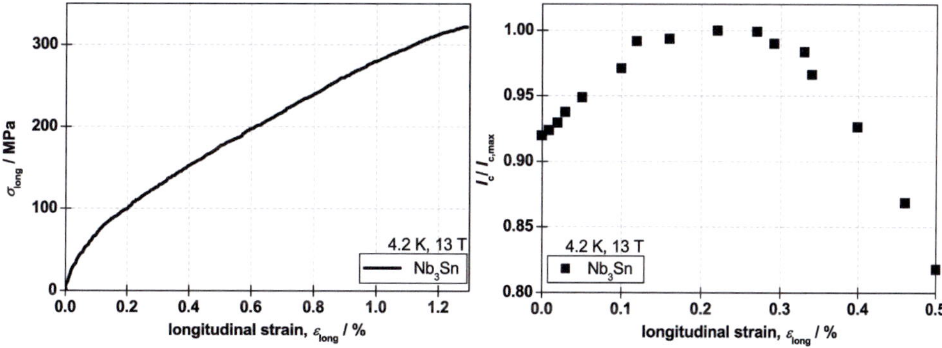

Figure 2.10: Mechanical properties of niobium-tin: Stress - strain dependency of Nb$_3$Sn (left), Influence of mechanical strain on the current carrying capabilities (right). Data from [Wei07].

to process and to wind it to coils. Winding to coils after heat treatment, "react & wind", is possible only if the Nb_3Sn is located in the neutral axis so that it doesn't experience tensile or compressive stresses. Commonly the "wind & react" or "wind, react & transfer" manufacturing methods are utilized. In both methods, the final heat treatment is done after Nb_3Sn is wound to the shape of the coils. Additionally the current carrying capabilities of Nb_3Sn strongly depend upon intrinsic and applied mechanical strain as shown on the right of **figure 2.10**. Mechanical strain and stress in Nb_3Sn coils has to be carefully controlled. Compressive strain is induced on Nb_3Sn coils due to compacting and heat treatment. This compressive "pre-strain" is partially compensated by the hoop stresses in operation increasing the current carrying capabilities.

Nb_3Sn is used in applications with higher magnetic fields on the conductor for example nuclear-magnetic-resonance (NMR) devices [ZIB92], high field magnets [HFPZ06] or some fusion magnets such as the toroidal field coils [SBOG10] and the central solenoid [LMB09] of ITER. Niobium-tin wires are commercially available from several companies around the world. The basic properties of the Nb_3Sn wires from different manufacturers used in the ITER project are shown in **table 2.2**.

Table 2.2: Basic properties of Nb_3Sn wires from Vacuumschmelze (VAC), IGC, Furukawa and Mitsubishi used for the central solenoid of ITER. Critical current density determined at 12 T, 4.2 K with a $0.1\,\mu V\,cm^{-1}$ criteria. Data from [TCRM99].

manufacturer	VAC	IGC	Furukawa	Furukawa	Mitsubishi
process	bronze	internal tin	bronze	bronze	internal tin
heat treatment	570 °C, 220 h 650 °C, 175 h	660 °C, 240 h	650 °C, 240 h	650 °C, 240 h	580 °C, 180 h 650 °C, 240 h
wire diameter	0.813 mm	0.811 mm	0.806 mm	0.809 mm	0.808 mm
twist pitch	9 mm	9 mm	17.7 mm	17.6 mm	16 mm
Cu : non-Cu ratio	1.45	1.47	1.51	1.48	1.56
J_c (non-Cu)	609 A mm^{-2}	681 A mm^{-2}	641 A mm^{-2}	664 A mm^{-2}	685 A mm^{-2}
J_e (total)	248.6 A mm^{-2}	275.7 A mm^{-2}	255.4 A mm^{-2}	267.7 A mm^{-2}	267.6 A mm^{-2}
n-value	35	34	36	38	27

2.3 High temperature superconductors

There are four technical high temperature superconductors: Rare-earth-barium-copper-oxide *RE*BCO, bismuth-strontium-calcium-copper-oxide (BSCCO) 2223 and 2212 and magnesium-diboride (MgB_2). The first two can be produced either as flat tapes or bulk material. The second two appear as round wires or bulk material. As a conductor for fusion-relevant magnets bulk materials are unsuitable. Therefore, bulk materials are not considered in the following work.

2.3.1 Bismuth-strontium-calcium-copper-oxide (BSCCO) 2223 tapes

Bismuth-strontium-calcium-copper-oxide ($Bi_2Sr_2Ca_2Cu_3O_y$ or BSCCO 2223) tapes are often referred to as first generation HTS material and are commonly manufactured using powder-in-tube manufacturing processes. In such processes the raw material (barium, strontium, calcium and copper) are milled, sintered and filled in small tubes. The tubes are closed and extruded. A bundle of such tubes is formed, which is again put in a tube and extruded. This is repeated several times. Through heat treatment in an oxygen rich atmosphere, oxygen has to deposited in the material. A high oxygen concentration is necessary for the superconducting properties. Therefore silver tubes which have a very high oxygen permeability are required [RWHM92, Mae12]. The silver in the tubes forms the matrix of the BSCCO 2223 tapes. The matrix couples the superconducting filaments and stabilizes the conductor electrically and thermally. In BSCCO 2223 a strict orientation of the grains is essential for high current carrying capabilities. This orientation is achieved by rolling the bundled tubes [Kom95, p. 92] to a flat conductor. The rolling provides the characteristic tape geometry which is shown in **figure 2.11**.

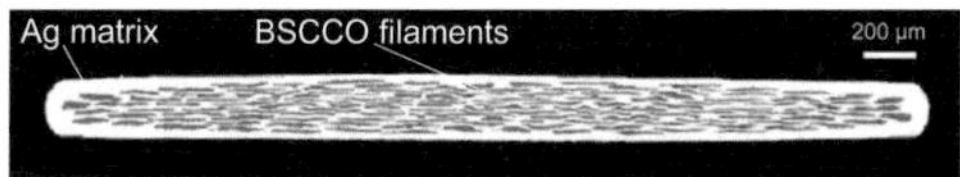

Figure 2.11: Cross section view of a BSCCO 2223 tape. The silver matrix is shown in white and the BSCCO filaments in gray. Picture from [GSR08].

Approximately half of a the cross sectional area of a BSCCO 2223 tape is its silver matrix. The silver has a significant impact on the raw material costs in BSCCO 2223 production [Hul03]. The production capacities of BSCCO tapes have been reduced over the last years. Today, they are commercially available from Sumitomo [Suma] and Innost [YSL05].

Mechanical properties of BSCCO 2223 tapes vary strongly. Depending on the lamination, the critical stress limits are in the 90 - 270 MPa range [Sumb, Sumc]. BSCCO 2223 tapes are very sensitive to shear stress, too. Shear stresses in the few MPa range can degrade the current carrying capabilities of BSCCO tapes significantly [BWG11]. This is shown in detail within the appendix, **section A.2**.

In high magnetic background fields the critical current densities of BSCCO 2223 tapes are lower compared with those of BSCCO 2212 wires and *RE*BCO tapes, especially in perpendicular orientation. Because of this, BSCCO 2223 tapes are commonly used in low and intermediate magnetic fields, e.g. in current leads [Bal02] and HTS power cables [MYI07]. For large dimensions and high magnetic background fields such as high field magnets or fusion magnets, BSCCO 2223 tapes are not used.

2.3.2 Bismuth-strontium-calcium-copper-oxide (BSCCO) 2212 round wires

Bismuth-strontium-calcium-copper-oxide ($Bi_2Sr_2Ca_1Cu_2O_y$ or BSCCO 2212) is a high temperature superconductor in a round wire shape. At high magnetic background fields ($\geq 15\,T$) the critical current density of BSCCO 2212 round wires is between that of Nb_3Sn and *RE*BCO tapes. The round shape of BSCCO 2212 wires is distinctly different from the other high temperature superconductors. With round conductors conventional cabling methods such as Rutherford cables or stranded cables are possible.

The BSCCO 2212 superconducting material is made from a powder. There are several manufacturing methods [MMM03]: powder-in-tube, different coating techniques such as thin film deposition and thick film deposition (PAIR method) or partial melting processes. BSCCO 2212 has the highest performance as a film or tape, since in this case an optimized texture of the grains is achieved. In contrast to BSCCO 2223, where a good 2D texture is essential, BSCCO 2212 can also carry high currents with a 1D texture along the conductor axis. This allows the fabrication of round BSCCO 2212 wires with a filament structure. Filament structures are necessary because the beneficial texture occurs preferentially at the Ag-sheath-filament interface. Therefore BSCCO 2212 round wire consists preferably of thin filaments which are embedded in a composite silver/silver-magnesium (AgMg) matrix. The same stacking and drawing techniques as for LTS multifilament wires can be used. A cross section view of an unreacted multifilament BSCCO 2212 wire is shown on the left of **figure 2.12**. To obtain the optimum superconducting properties the material needs to be reacted by a sophisticated heat treatment which involves a partial melting of the superconducting core in a limited time and temperature interval at quite high temperatures (around 890°C). This is the critical step in the manufacturing process of this superconductor. Depending on the powder composition and the precursor used, the material becomes inhomogeneous. During the heat treatment oxygen exchange with the surrounding atmosphere takes place. Formation of intergrowths between the filaments and compaction due to the partial melting can generate large pores in the wires. This is shown on the right of **figure 2.12** for a conductor manufactured by Oxford Superconducting Technologies (OST) [BDF12b].

The void fraction of the wire increases during heat treatment. Filaments can be isolated locally and the current carrying capability of the whole conductor is reduced. In the production of long conductors this is dominant as the gas generated can only escape at the ends [MKI11]. With improved production methods, such as heat treatment under higher pressure [Liu11] and swaging [JMH13], the wires are densified and this effect can be reduced and controlled. Optimization of the processing for high currents leads to the formation of strong intergrowths between the filaments. As a consequence the filaments are coupled and the magnetic component of the AC losses are worse than in the ideal case of isolated filaments [BDF12b].

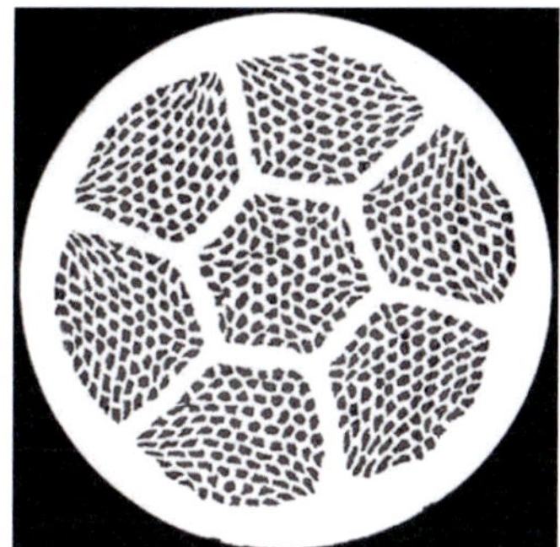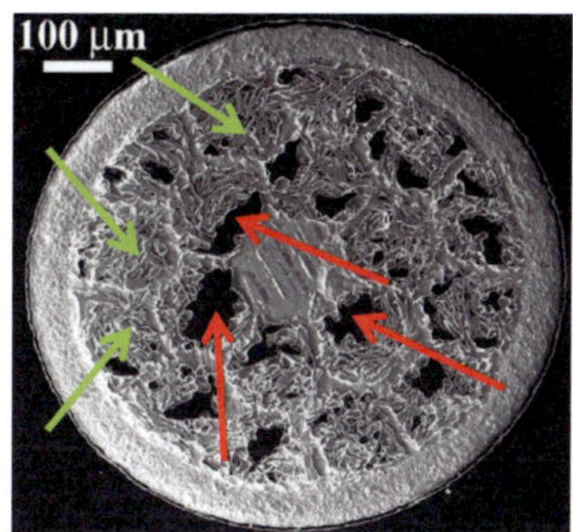

Figure 2.12: Bismuth-strontium-calcium-copper-oxide 2212 round wires: Cross section view of an unreacted 91x7 filament wire with a diameter of 0.8 mm (left). Picture from [MMM03]. ©2003 IEEE. BSCCO 2212 wire after heat treatment (right). Well-connected filaments (green arrows) and large pores (red arrows) are visible. Reprinted with permission from [BTK08]. ©2008 American Institute of Physics.

Similar to BSCCO 2223 tapes, there is a a significant fraction of silver in BSCCO 2212 round wires, increasing the raw material costs and reducing the mechanical properties. The critical strain limits are less than 0.4 % in compression and tensile [TTT95], the critical stress limits are in the 26 - 43 MPa range [PPL99]. Mechanical cracks usually originate from voids within the wires. Therefore, it can be assumed that reductions of the voids will also improve the mechanical properties [PPL99].

BSCCO 2212 round wires are manufactured only by Oxford Superconducting Technologies (OST) in large quantities. At present, they are not feasible as a conductor in large magnets, further improvements of the manufacturing processes are necessary to suppress the forming of pores completely.

2.3.3 Rare-earth-barium-copper-oxide (*REBCO*) tapes

Rare-earth-barium-copper-oxide (*REBCO*) high temperature superconductors are crystalline chemical compounds with formulas of $Rare\text{-}Earth_1Ba_2Cu_3O_{7\text{-}x}$. They crystallize in a defective perovskite structure in different layers. The macroscopic superconducting properties depend on the saturation with oxygen and the alignment of the crystal grain boundaries. With low oxygen concentration ($0.55 \leq x \leq 1$) the copper layers are oxygen depleted and the material behaves as an electrical insulator [Oak96]. Superconductivity is observed if the copper layers are partially saturated with oxygen ($x < 0.55$). The critical temperature depends on the oxygen concentration. A maximal critical temperature ($T_c > 90$ K) is achieved with oxygen concentrations in the range of $0 < x \leq 0.2$ [Oak96]. Superconductivity is very anisotropic and there is a low coherence length ($\xi_c \approx 0.4$ nm to $\xi_{ab} \approx 2$ nm [KFC06, p. 16]) observed compared with low temperature superconductors ($\xi_{Nb_3Sn} \approx 3.6\,\mathrm{nm}$ [God06] and $\xi_{NbTi} \approx 5\,\mathrm{nm}$ [LZL08]). In *RE*BCO, the superconducting state is highly susceptible to disruptions in the crystalline structure as found at the boundary of differently oriented grains [Alb04]. A single crystal or very good alignment between the grains is critical for the superconducting properties as the current carrying capabilities are strongly influenced by the angles between the grains [GMP04]. The superconductivity of these materials was discovered in 1987 using yttrium as rare earth component therefore these superconductors are commonly referred to as yttrium-barium-copper-oxide (YBCO). As these superconductors are the most promising materials for future fusion magnets, *RE*BCO is covered in detail in the following.

2.3.3.1 Composition and manufacturing of *RE*BCO

Rare-earth-barium-copper-oxide (*RE*BCO) can either be grown as bulk crystals or it can be deposited as a thin film on metal substrate tapes which is called a coated conductor (cc). In general, large polycrystalline bulk *RE*BCO has a lower critical current density and is unsuitable for fusion magnets. Instead, bulk *RE*BCO is usually used for magnetic levitation [MPC03] and magnetic bearings [WFDR12]. In this work only the deposited *RE*BCO films, the coated conductors, are of interest. *RE*BCO and YBCO are in the following used as a synonym for the coated conductors tapes.

Coated conductors have a flat tape geometry. The tapes are of a thickness of $50\,\mu m$ to $200\,\mu m$, a width of up to $40\,mm$ and are commercially available in up to $1.4\,km$ pieces (in 2011) [Sel11c]. They consist of a steel substrate (Hastelloy, stainless steel or nickel alloys) of $50\,\mu m$ to $120\,\mu m$ which is coated with several metal-oxide buffer layers. The buffer layers prevent diffusion between substrate and *RE*BCO layer. Additionally, the buffers compensate the lattice mismatch, allowing the growth of homogeneous *RE*BCO layers. For strong superconducting properties, high critical current densities and critical magnetic fields, a good alignment of the grains is needed on a long length scale. This alignment is achieved through texture, either with a bi-axial textured substrate, using rolling assisted bi-axial textured substrate (RABITS) or alternating beam assisted deposition (ABAD) processes, or by a bi-axial texture in the buffer layers, with ion beam assisted depositioning (IBAD). The texture determines the orientation of the grains and reduces the angles between the grain boundaries. This is critical for high performance tapes. The uppermost buffer layer is coated with a very thin film of rare-earth-barium-copper oxide, the superconducting layer. It is $1\,\mu m$ to $3\,\mu m$ thick and it carries the complete current of the whole tape in superconducting state. The current density in the superconducting layer is very high ($10\,kA\,mm^{-2}$ range). It is covered with a very thin silver, silver-gold or gold cap layer of a few μm to improve the distribution of the current and the heat, stabilizing the superconductor thermally and electrically.

For additional electrical stabilization, there are versions of *RE*BCO tapes available which

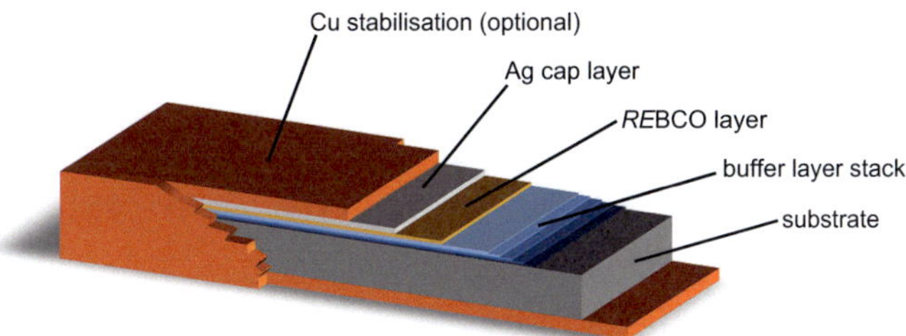

Figure 2.13: Layout of *RE*BCO coated conductors. The layers are scaled arbitrarily.

are electro-plated with copper of variable thickness (from $20\,\mu m$ to $100\,\mu m$). The layout of *RE*BCO coated conductors is shown schematically in **figure 2.13**.

Manufacturers of *RE*BCO use various compositions and different rare earth elements. Yttrium is mainly used by American Superconductor (AMSC), Bruker EHTS and SuperPower while Fujikara, ISTEC, Sumitomo Electric and Furukawa use gadolinium. Samarium is used by KERI and dysprosium by Theva. There are also differences in layer thicknesses or the used thin film depositioning method. Coated conductors are manufactured with pulsed laser depositioning (PLD) by Fujikura [Fuj], ISTEC, Bruker EST [Die06] and Sumitomo Electric [YOIS11] and with metal organic chemical vapor deposition (MOCVD) by SuperPower [HRB11]. AMSC [LRT09] and Furukawa [Die06] utilize metal organic deposition (MOD). KERI uses an evaporation drum in dual chamber (EDDC) process for high performance tapes and reactive co-evaporation by deposition and reaction (RCE-DR) for high tape output [Oh11]. Inclined substrate deposition (ISD) is utilized by Theva [PNH05]. These differences in the substrate material, the thickness of the layers and the manufacturing processes result in tapes with varying mechanical, thermal and electrical properties. The constitution and composition of *RE*BCO tapes of major manufacturers is given in **table 2.3**.

2.3.3.2 Properties of *RE*BCO

Rare-earth-barium-copper-oxide (*RE*BCO) tapes are products which are actively being developed. Many improvements of *RE*BCO are challenging to realize in commercial mass production, thus short laboratory samples are usually superior compared with the commercially available material.

Mechanical properties
Depending on the substrate material, the tape is resistant to longitudinal tensile load due to a high Young's modulus and elastic deformation up to nearly $0.8\,\%$ tensile strain. Longitudinal tensile strain has a low influence on the current carrying capabilities as shown on the left of **figure 2.14** compared with other superconductor materials. More than $0.5\,\%$ longitudinal strain can be applied in tape direction with less than $5\,\%$ degradation of the current carrying capabilities (stainless steel laminated tapes, data from [Lam06][GSR08, p. 40]). In addition, this depends

Table 2.3: Composition of commercially available coated conductor tapes from different manufacturers. Data from: SuperPower [Supd, All]; AMSC [LSB10, FRRS08]; Bruker [Bru12, Uso10, Uso09] and others [YOIS11].

	Fujikura	ISTEC	Sumitomo	Furukawa	SuperPower SCS12050	AMSC	Bruker EHTS
process	PLD	PLD	PLD	PLD	MOVCD	MOD	PLD
texture	IBAD	IBAD	RABITS	IBAD	IBAD	RABITS	ABAD
substrate / μm	100	100	120	100	50	75	50
buffer / μm	1.18	0.63	0.51	0.495	1	0.225	0.155
REBCO / μm	2	2	3	0.85	1	1.2	1-3
Ag layers / μm	20	20	10	20	1	3	0.2
Cu / μm	-	60	40	100	40	50	40
Cu / %	0.4	33.1	23.5	45.3	42.9	38.8	42.9
Ni / %	44.7	30.1	65.7	24.9	29.2	55.1	42.2
Mo / %	12.2-13.8	8.2-9.3	-	6.8-7.7	8.0-9.0	-	-
Cr / %	11.8-13.4	7.9-9.0	-	6.6-7.5	7.7-8.7	-	11.4
Fe / %	3.2-5.7	2.2-3.8	-	1.8-3.2	2.1-3.7	-	-
W / %	2.4-3.7	1.6-2.5	3.5	1.4-2.0	16.6-2.4	2.9	-
Ag / %	16.2	11.0	5.8	9.0	2.1	2.3	0.1-0.2
Y / %	-	-	-	<0.1	0.1	0.1	0.1-0.2
Ba / %	0.2	0.2	0.3	<0.1	0.2	0.1	0.2-0.3
other / %	1.9	1.0	1.3	0.4	1.7	0.7	2.2-2.5

not only on the strength of the strain, but also on the orientation. At angles between the strain direction and the direction of the tapes of 0° and 90°, the reduction is maximal. At angles of 45°, mechanical strain hardly influences the critical currents. The anisotropic strain effect of REBCO tapes is explained in detail within the appendix, **section A.1**. In applications, the longitudinal (parallel to the tape surface) critical tensile stress limits of REBCO tapes are in the 500 - 700 MPa range [Haz10, Bru12, Supd, Sel11b].

Up to 150 MPa of compressive transverse stress (perpendicular to the tape surface) can be applied on REBCO tapes from AMSC without measurable degradation [CETX07] as shown in black on the right of **figure 2.14**.

On SuperPower tapes up to 240 MPa of compressive hydrostatic pressure is possible [YLJ12]. Transverse (perpendicular to the tape surface) tensile stresses strongly influence the tapes. For example, the current carrying capabilities of REBCO tapes from AMSC are reduced by 0.38 % at transverse tensile stress of 24.8 MPa. This reduction is fully reversible. However, at above

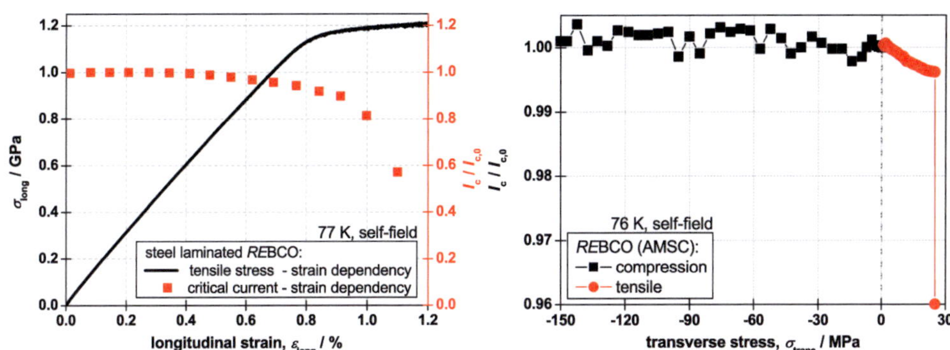

Figure 2.14: Mechanical properties of *RE*BCO tapes. On the left: Longitudinal tensile stress (σ_{long}), strain ($\varepsilon_{\text{long}}$) dependency and dependency of the current carrying capabilities on tensile strain stainless steel laminated tapes. Data from [Lam06]. On the right: Influence of transverse compressive stresses and transverse tensile stresses (σ_{trans}) on the current carrying capabilities of AMSC tapes. Compressive components are shown in black, data from [CETX07], while tensile components are shown in red, data from [LECS07].

25 MPa of transverse tensile stress, delamination in the *RE*BCO layers occurs [LECS07]. The tapes are damaged irreversibly and the current carrying capabilities are strongly and permanently reduced, as shown in red on the right of **figure 2.14**. In applications it is therefore important to carefully control transverse tensile stresses.

Rare-earth-barium-copper-oxide (*RE*BCO) tapes can be bent to small radii. Bending radii of 60 mm are possible with less than 1 % degradation of the current carrying capabilities [GSR08, p. 28] if the superconducting layer is on the compressive side. Bending the tapes to lower radii drastically reduces their critical current. However, the bending remains fully reversible down to radii of 6 mm [SMS11]. Shear stresses, which can occur by twisting *RE*BCO tapes, neither damage the tapes nor influence their current carrying capabilities, as investigated within the appendix, **section A.2** for shear stresses up to 48.7 MPa on SuperPower tapes. Some manufacturers provide *RE*BCO tapes with further improved mechanical properties. This is achieved with thicker substrates [Supe] or by laminating the tapes with additional structural material [Supa, type: stainless steel] (as shown on the left of figure **figure 2.14**).

Current carrying capabilities

Critical current densities within the superconducting *RE*BCO layer of several 10 kA mm^{-2} have been achieved[a]. In an *RE*BCO tape the fraction of *RE*BCO is small as the superconducting layer is only of a few µm. At present, *RE*BCO tapes of long lengths (> 300 m) with critical currents in the 300 - 425 A/cm-width range are available [Ame, Leh13]. On short samples, critical currents of more than 1000 A/cm-width have been achieved [Oh11]. The current carrying capabilities strongly vary depending on the manufacturer and the quality of the material [Bru12, Supd, Oh11].

---

[a]Calculated using the data from [Oh11].

For high currents per width in long samples (in the range of several hundreds of meters) the stability of the manufacturing process is essential. There are variations in critical current along the tape length as shown on the example of SuperPower tapes from 2011 in **figure 2.15**. To achieve high quality products, manufacturers are trying to minimize these variations [Haz10, SD10, Sel11b].

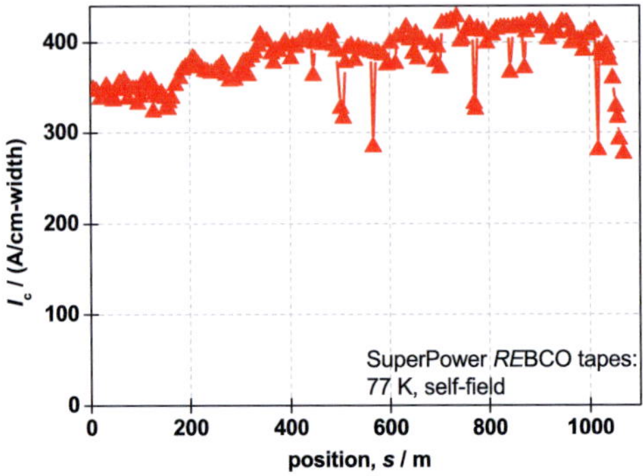

Figure 2.15: Variations in the critical current of SuperPower REBCO along the length of the tape. Data from [Sel11b].

Magnetic field dependence

Rare-earth-barium-copper-oxide (REBCO) tapes are highly anisotropic materials. The reduction of the critical current due to magnetic background fields not only depends on the strength of the magnetic field but also on the field orientation. For a given field strength, magnetic fields perpendicular to the tape direction commonly have the largest impact on the critical current. Fields parallel to the tape result in the lowest reduction of the current carrying capabilities. Thus the dependence of the critical current on magnetic background fields is commonly given for REBCO for these two orientations. As shown in **figure 2.7**, the magnetic field perpendicular to the tape surface $j_c \perp$ is far lower than magnetic field parallel to the tape surface $j_c \parallel$ in REBCO tapes. This is referred to as the anisotropy of REBCO tapes. Manufacturers are trying to improve the in-field performance of REBCO and reducing the anisotropy. With a high anisotropy, areas with perpendicular field orientation can be limiting factors in applications. Such improvements can be achieved by increasing the number of pinning centers in the superconducting layer. This has been achieved by the manufacturer SuperPower, by zirconium (Zr) doping [Sel11c] of their tapes. The zirconium atoms act as additional pinning centers and increase the pinning force. This conductor is called an "advanced pinning" conductor. In **figure 2.16**, the dependence of the critical current density on the angle of applied magnetic fields is shown for undoped and Zr doped SuperPower REBCO tapes. At field orientations other than parallel (90° and 270°) the improvements of the Zr doped conductor are significant. Due to the non regular shape of the curve, considering only parallel and perpendicular fields is not applicable any more. Such a simplification is not suitable

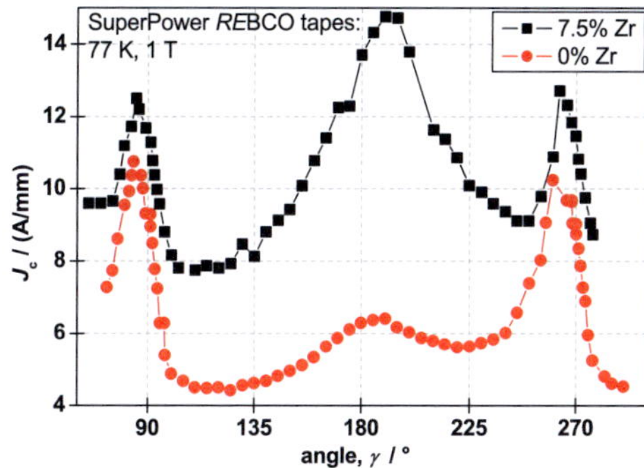

Figure 2.16: Increase of in-field performance and reduction of the anisotropy of SuperPower *RE*BCO tapes due to zirconium (Zr) doping. Data from [Sel11c].

Table 2.4: Properties of commercially available *RE*BCO tapes. Data from: AMSC [Mae12]; Bruker EST [Bru12]; Fujikura [Fuj]; Sumitomo [OYM10]; KERI [Oh11]; ISTEC [Haw06] and SuperPower [Supd].

manufacturer	product	width / mm	I_c / A/cm-width	σ_{long} / MPa	stabilization	process	length / m
AMSC	copper 4.8	4.8	≥ 200	150	Cu	RABITS - MOD	
	copper 12	12	≥ 200	150	Cu		
	brass 4.8	4.25	≥ 425	200	brass		
Bruker EST	YHT	4	338	650	Cu	ABAD - PLD	2000
Fujikura		10	572		Cu	IBAD - PLD	
Sumitomo	CC1	2	250	500	Cu	IBAD - PLD	
KERI	high current	4	1000		-	EDDC	16
	high output	4	355		-	RCE-DR	920
ISTEC		10	245		-	IBAD - PLD	212
SuperPower	SF	4, 12	> 250	550	-	IBAD - PLD	1000
	SCS	4, 12	> 250	550	Cu		

for state-of-the-art *RE*BCO (e.g. advanced pinning) tapes. A characterization considering the magnetic field strength and the angle of the field is required instead.

Table 2.4 summarizes the properties of commercially available *RE*BCO tapes from different manufacturers.

2.3.3.3 Production of *RE*BCO

Production of *RE*BCO tapes in long lengths is a challenging technology. At present several companies are producing *RE*BCO commercially. The main supplier of *RE*BCO tapes is Super-Power. As shown on the left of **figure 2.17** SuperPower's production output has increased from short laboratory samples in 2002 (90 A m) to a mature product available in several hundreds of meter pieces in 2009 (300.33 kA m) [Sel11c]. Several other companies are also working on

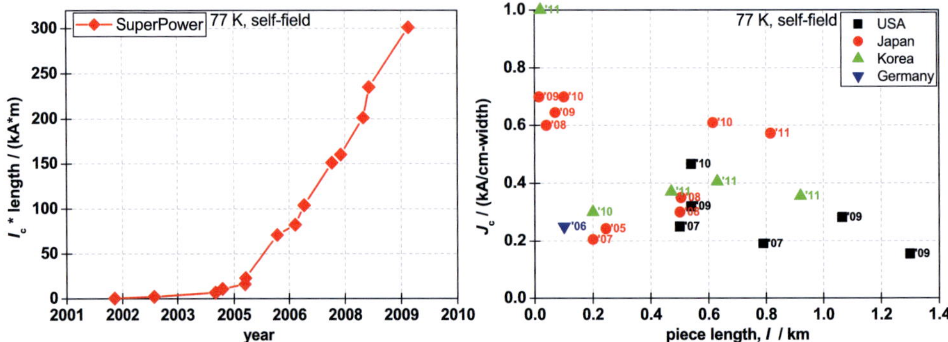

Figure 2.17: SuperPower's *RE*BCO production from 2002 to 2009. Data from [Sel11c] (left). Critical current density vs. unit length of *RE*BCO tape production of different countries. Data from [Oh11] (right).

*RE*BCO mass production. In the United States (SuperPower and American Superconductor), in Japan (Sumitomo, Fujikura, Furukawa and ISTEC), in Korea (KERI) and in Germany (Bruker EST and Theva) are involved in *RE*BCO tape production. As shown on the right of **figure 2.17** there have been large improvements in the performance of the tapes and in the piece lengths in the last years. At present several companies are able to manufacture tapes with several 100 A/cm−width in pieces of several hundreds of meters [Oh11]. It is to be expected that these companies will further increase their production capabilities [Uso09] and will also enter the commercial *RE*BCO tape market [Oh11].

Due to the complex manufacturing process *RE*BCO coated conductor tapes are expensive. *RE*BCO tapes from SuperPower cost about 225 $/(kAm) in 2011 [Lor11]. For large scale applications, a reduction to $1/10^{\text{th}}$ of this price is necessary to be competitive to low temperature superconductors or conventional conductors such as copper [Mae12].

2.3.3.4 Outlook on production and performance of *RE*BCO tapes

Research and development of *RE*BCO tapes are actively pursued by the manufactures and research institutions mainly in the United States, Europe and Asia. Manufacturers are increasing their production capabilities and improving the performance of the tapes as shown in **figure 2.18**

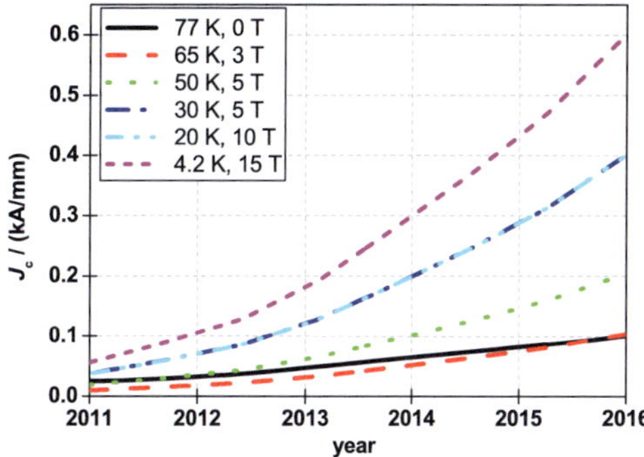

Figure 2.18: Outlook on *RE*BCO tape performance as seen by the manufacturer SuperPower. Data from [Sel11a].

for *RE*BCO tapes from SuperPower [Sel11a]. Through production in larger lengths the manufacturers are trying to increase the cost efficiency and reduce the manufacturing costs of the tapes [Uso09]. Improving the performance of the tapes also has a positive effect on price performance ratio \$/(kAm). Manufacturers are trying to increase the current carrying capabilities of *RE*BCO tapes by increasing the thickness of the superconducting layers [Sel11c, Oh11]. Nano structures (nano rods) in the superconducting layers have the potential to provide additional pinning centers [Sel11b]. Also, as nano structures can be placed systematically, they can help to improve the tape homogeneity. These pinning centers further increase the pinning force and improve the in field performance of *RE*BCO tapes. SuperPower, which is at present among the major commercial manufacturer of *RE*BCO tapes, sees great potential for further increasing the performance of their products.

Such improvements and a significant increase in production volume are necessary for *RE*BCO tapes to become competitive and to establish large scale application markets e.g. superconducting fusion magnets.

2.4 Comparison of superconductors and their usability for fusion magnets

In **table 2.5**, the properties of technical superconductors NbTi, Nb$_3$Sn, *RE*BCO, BSCCO 2212 round wires and BSCCO 2223 tapes are summarized and compared. Due to their weak mechanical properties and high silver contents BSCCO 2212 round wires and BSCCO 2223 tapes are not suitable conductors for fusion magnets. In the high field areas, e.g. the toroidal field coils of fusion magnets, Nb$_3$Sn and *RE*BCO tapes are applicable. *RE*BCO has a higher critical current density at high fields (15 T and above) and much better mechanical properties, making *RE*BCO a promising alternative to Nb$_3$Sn for the magnets of future fusion plants. Due to *REBCO*'s critical temperature in the 90 K range, operation with higher temperature margins or at higher cable temperatures are possible.

Table 2.5: Comparison of technical superconductors: NbTi, Nb$_3$Sn, *REBCO*, BSCCO 2212 round wires and BSCCO 2223 tapes.

	NbTi	Nb$_3$Sn	BSCCO 2223	BSCCO 2212	*REBCO*
in field usability	medium fields	high fields	low fields	very high fields	very high fields
mechanical properties	very good - ductile	bad - brittle	depend on lamination	bad	very good
strain range	high	low	medium	medium	high
$J_c(\varepsilon)$ dependence	low	high	medium	medium	low
raw material costs	low	low	high	high	low
manufacturing	simple	medium	medium	complex	very complex
availability	very good	very good	bad	bad	medium
usability for fusion	medium field areas (PF coils)	high field areas (TF, CS coils)	-	-	high field areas (TF, CS coils)

3 Superconducting fusion magnets

3.1 Introduction to fusion

Nuclear "fusion is the process which heats the Sun and all other stars. It is one of the most promising options to generate sustainable, carbon-free energy in the future" [Cul12].

Two or more light atoms are combined into heavier atoms in fusion reactions. Light atoms with less the 26 protons (up to iron) have a low binding energy per nucleon, they are energetically not ideal. The sum of the nuclear binding energies of the light reactants is lower than the binding energy of the reaction products. By "fusing" them together the difference of the nuclear binding energies is dissipated. Fusion reactions, similar to other nuclear reactions, are suppressed by the Coulomb Barrier. The Coulomb Barrier is the energy barrier of the repellent Coulomb potential ($\sim \frac{1}{r}$) between positively charged nuclei, which makes matter stable in solid, liquid and gaseous aggregate phases. Hot plasmas of light atoms with thermal energies in the million degree range are necessary to overcome the energy barrier at reasonable atomic cross-sections allowing nuclear fusion.

The main fusion reaction in the Sun and in stars is the proton-proton reaction. On earth, due to the highest fusion power density, the most commonly used fusion reaction is the fusion of heavy hydrogen, deuterium (H_2), with super heavy hydrogen, tritium (H_3), to helium (He_4) and a neutron (n). For this reaction, plasma temperatures in the 10^8 kelvin regions are ideal to ignite the tritium and deuterium fusion processes [EFD]. Fusion reactions have an enormous fuel efficiency. About 0.38 % of the total mass of the fuel is dissipated as energy in the case of the fusion of tritium and deuterium. Over 80 % of this energy is available as kinetic energy of the neutron as shown schematically in **figure 3.1**. Additionally deuterium and tritium are abundant on earth [HB01] which makes nuclear fusion a desirable technology for the generation of electrical power for future generations [Oak].

3.1.1 Magnetic confinement

In controlled fusion reactions, temperatures in the 10^8 K region have to be maintained over long time periods [EFD]. A contact-less confinement of an extremely hot plasma is necessary. The confinement of the plasma is a central part of controlled fusion. As the charged plasma particles are influenced by magnetic fields, magnets have been the most common confinement method.

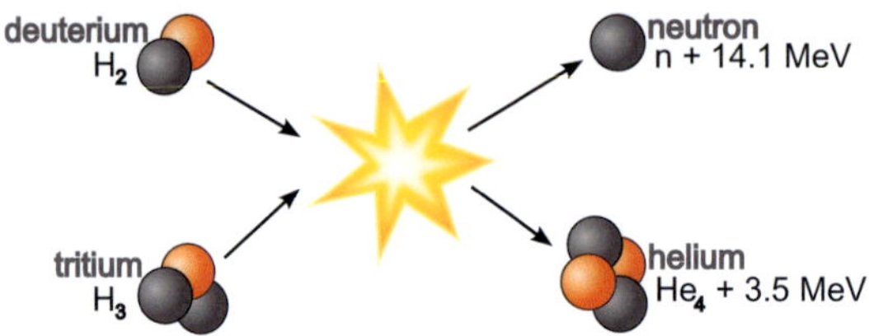

Figure 3.1: Main fusion process used on earth. Deuterium and tritium are fused to helium and a neutron. More than 80 % of the energy is dissipated as kinetic energy of the neutron. Picture from [Wik07].

In the early years of controlled fusion, experiments with various magnet configurations, e.g. the perhapstron [Phi83] and magnetic mirrors have been conducted [Rag11]. From 1975 the Tokamak magnet system and the Stellarator magnet system, which both generate toroidal helical magnetic fields, have become the dominant methods of plasma confinement [SBG01].

Tokamak

Tokamak is a transliteration from Russian, where it is an acronym of either "toroidal chamber with magnetic coils" or "toroidal chamber with axial magnetic field". The magnet system of a Tokamak type plasma confinement system (Tokamak fusion reactor) consists mainly of three different kind of coils. The toroidal field (TF) coils generate a doughnut shaped magnetic field around the torus. The primary function of these coils is to confine the charged plasma particles [ITE]. In addition, there are the poloidal field (PF) coils which stabilize and shape the plasma and pinch the particles away from the vacuum vessel. For a helical component which is required to complete the magnetic confinement, a plasma current is induced. The helical magnetic field of this current combined with the magnetic fields of the toroidal field coils and the poloidal field coils form the "magnetic bottle". The plasma current is induced through the central solenoid (CS). The central solenoid is ramped up and down and acts as the primary winding of a transformer. In the left part of **figure 3.2** a Tokamak magnet system with its toroidal field coils (red), poloidal field coils (dark green), the central solenoid (gray), the vacuum vessel (light blue) and the plasma (yellow) is shown schematically.

Stellarator

Stellarator is derived from the Latin word "Stella" which means star. The magnet system of a Stellarator type fusion reactor consists mainly of two different kinds of coils. Comparable to the Tokamak type magnet system there also are toroidal field coils (TF) which generate the toroidal field. The helical component which is needed to complete the magnetic confinement is generated directly by helical field coils (HF) which are helically wound around the vacuum vessel. In contrast to Tokamaks, a plasma current is not necessary in Stellarators, no coils have to be ramped and steady state fusion is possible intrinsically. Key components of a Stellarator magnet system are shown schematically on the right of **figure 3.2**. The toroidal field coils are shown in red, the helical field coils in green, the vacuum vessel in blue and the plasma in yellow. The helical field coils of Stellarator type magnet systems are very large non-planar coils which are extremely difficult to design and to manufacture.

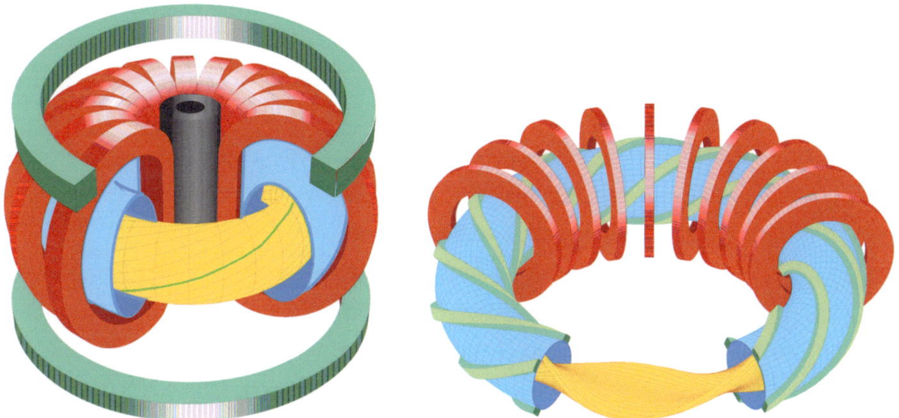

Figure 3.2: Schematic drawing of magnet systems for magnetic confinement of plasmas: Tokamak magnet system on the left and Stellarator magnet system on the right. Toroidal field coils are shown in red, poloidal field coils in dark green, the central solenoid in gray and helical field coils in light green. Vacuum vessels are shown in blue and plasmas in yellow. Pictures from [Max12b] and [Max12a].

There have been more Tokamak type fusion experiments than Stellarators and at present the Tokamak type reactors are favored as prototypes for future fusion power plants. Thus this work mainly concentrates on the Tokamak type fusion reactors.

3.1.2 Development of magnets for controlled fusion

After the second world war there was an increasing interest in fusion all around the world. Fusion experiments were conducted in several machines in the Soviet Union [Vel08], the United States [Sti98], Japan [YMS81], the United Kingdom [PCC81], France [Equ78], Italy [ABB84] and Germany [Pul85]. These fusion machines used resistive magnets made from copper to generate the magnetic fields needed for the containment of the plasma. From these experiments it became clear certain plasma pressures and very high temperatures are required to reach ignition. These can be achieved only by long containment times and high plasma volumes. This results in large and expensive machines which are often built in international collaborations.

The magnets for the containment of the plasma are a key component in fusion experiments. Beginning with the magnetic mirror experiment "Baseball I", superconductors have been used in fusion magnets to deliver high magnetic fields with low losses. Niobium alloys are the preferred low temperature superconductor materials. Commonly used are niobium-tin and niobium-titanium (see **subsection 2.2**). Superconducting magnets have been steadily improved. Their cables evolved from single strand niobium-zirconium superconducting wires (Baseball I) [HLCB82] to monolithic multistrand cables and to cable-in-conduit conductors (CICC) (see **subsection 3.1.3**). These improvements paved the way for large efficient magnets which are essential for big fusion machines. At present Tore Supra, EAST and KSTAR are operational Tokamak type fusion experiment using superconducting magnets, while SST-1 and the International Thermonuclear Experimental Reactor (ITER) are experiments which are under construction. In **table 3.1** the development of large superconducting magnets is shown chronologically. Superconducting magnetic mirrors, Tokamak type experiments, Stellarator type experiments and important superconducting demonstration coils are listed.

Superconductivity and magnetic plasma confinement are linked. In the Joint European Torus (JET) which is at present the largest fusion experiment in operation, resistive copper magnets are used to contain the plasma. The total energy required to power JET's magnets, which is above 1 GW, can only be delivered to the machine in short pulses of 10 s to 30 s during plasma discharges. As the amount of power which can be drawn from the grid is limited, large energy storage flywheels are used to generate the pulses [CEA07]. Obviously this is not an adequate solution for a large continuously operating reactor and therefore not an option for a fusion power plant. Superconductivity enables the construction of magnets which are large and powerful enough to contain the plasma volumes needed for sustained fusion reactions without requiring enormous amounts of energy. Therefore all large fusion experiments, which are planned or are under construction, as SST-1, W7-X, JT60-SA and ITER use superconducting magnets.

3.1.3 State-of-the-art of fusion magnet technology

Modern fusion magnets use liquid helium cooled low temperature superconductors, niobium-titanium or niobium-tin depending on the magnetic background field.

3.1.3.1 Cable-in-conduit conductors

In fusion magnets, the superconductors are commonly assembled into force-flow cooled cable-in-conduit conductors (CICC). In such cables, superconducting wires and copper wires are entwisted with various methods such as Roebel cabling, Rutherford cabling, braiding or stranding.

In high current cable-in-conduit conductors stranding is mainly used. Sometimes not all strands are stranded at once, the stranding is done in several stages. Sub-cables are formed by stranding thin wires which are then entwisted themselves. This allows low void fractions and complete transposition of all wires [Kom95, p. 64 f.]. In magnets the superconducting wires are exposed to varying magnetic fields. Without any transposition, the area surrounded by superconducting

Table 3.1: Development of superconducting magnets for plasma confinement. Different types of superconducting fusion magnets are listed: Type "C" are demonstration coils, type "M" are magnetic mirrors, type "T" are Tokamaks and type "S" are Stellerators. Fusion magnet systems in blue are in operation while magnet system in red are under construction.

year	name	type	country	B_{max} / T	I_{op} / kA	energy / MJ	super-conductor	cable type	cooling	references
1965	Baseball I	M	US	3.8	0.23		NbZr	single strand	pool	[HLCB82]
1969	SUMMA	M	US	10.3	0.3	10.3	NbTi, Nb_3Sn	untwisted	pool	[RSN73, Luc69]
1970	Baseball II	M	US	7.5	2.4	17	NbTi	untwisted	pool	[HNC71, Hen78]
1970	LIN-5	M	Ru	5.8	1	<1	NbZr		pool	[Bru06]
1971	IMP	M	US	8.5	0.82	2.24	Nb_3Sn	monolithic	pool	[ECB71]
1971	Bumpy Torus	M	US				NbTi			[RRG73, Hen78]
1973	LIN-5B	M	Ru	8.3	6	2.8	NbTi		pool	[Bru06]
1978	T-7	T	Ru	4.8	6	20	NbTi	flat cable	forced flow	[NSC77]
1979	TESPE	T	D	7	7	8.8	NbTi	flat cable	pool	[TMP85]
1981	MFTF	M	US	7.68	5.8	409	NbTi	monolithic	pool	[DHH79]
1986	TRIAM-1M	T	J	11	6.2	76	Nb_3Sn	monolithic	pool	[JIN97]
1984	LCT	C	US, J, Eu, Ch	8	10-18	758	NbTi, Nb_3Sn	various	pool, f. flow	[Bru06]
1988	Tore Supra	T	F	9	1.4	600	NbTi	monolithic	pool	[ABC81]
1989	T-15	T	Ru	9.3	5.6	790	Nb_3Sn	flat cable	forced flow	[Che93]
1990	DPC	C	US, J	9.5-12	10-30	≈ 7	Nb_3Sn	CICC	forced flow	[SKN93, Nak88, SPT91]
1994	POLO	C	F, D	1.63	15	2	NbTi	CICC	pool, f. flow	[DFH97]
1998	LHD	S	J	9.2	17.3	1700	NbTi	CICC	pool, f. flow	[SM02]
2006	EAST	T	CN	6.5	16.4	TF: 390	TF: NbTi	CICC	forced flow	[CPCW08]
2008	KSTAR	T	KO	7.2	35.2	TF: 470	TF: Nb_3Sn	CICC	forced flow	[KPP05]
2012	SST-1	T	IND	5.1	10	TF: 56	NbTi	CICC	forced flow	[Sax00]
2014	W7-X	S	D	6.2	16	620	NbTi	CICC	forced flow	[Sap00]
2016	JT-60SA	T	J, Eu	TF: 5.65	TF: 25.7	TF: 1006	TF: NbTi	CICC	forced flow	[IBK10, YTK10]
2019	ITER	T	ITER Org.	TF: 11.8	TF: 68	TF: 41000	TF: Nb_3Sn	CICC	forced flow	[ITE]

wires which is penetrated by the field can be very large in a magnet of fusion relevant size. Large currents, commonly referred to as coupling currents, are induced. These currents reduce the transport currents which can be carried by such a superconducting cable. This can also result in low coupling and alternating-current (AC) losses. In **figure 3.3** this is shown schematically for a cable consisting of two superconducting wires.

Transposition of the wires or strands is therefore essential in fusion relevant superconductor cables. To minimize the coupling currents and the AC losses, the length of the transposition, commonly referred to as the "twist pitch" of the cable should be as low as possible. However, short twist pitches increase the necessary amount of superconductor per cable length, which has to be balanced against the AC losses in finding the ideal twist pitch of the cable. Copper wires are commonly included to provide additional electrical and thermal stabilization. The assembled cables are equipped with jackets of structural material for mechanical stabilization. In forced-flow cooled CICC cables the cooling agent is forced through the voids between the wires due to the pressure gradient between the start and the end of the cable. In some CICC a central cooling channel is included to reduce the flow resistance. Cable-in-conduit conductors using different cabling methods are shown in **figure 3.4**.

Cable-in-conduit conductors [HM75] have been used in fusion magnets since the Large Coil Task (LCT) in 1984. In ITER, the largest fusion experiment, the mechanical stabilization provided by the jacket of the CICC is not sufficient in the toroidal field coils, in which the Lorentz forces are highest. Instead, the jacket of the CICC is rather thin but the whole cable is embedded in radial plates of structural material [KSSP00]. The radial plates stabilize the toroidal field coils of ITER mechanically but they also reduce the overall current density of the coils. In **figure 3.5** the cable-in-conduit conductors of the ITER magnet system are shown. The conductor of the poloidal field coils and central solenoid is shown on the left of that figure and the conductor for the toroidal field coils on the right.

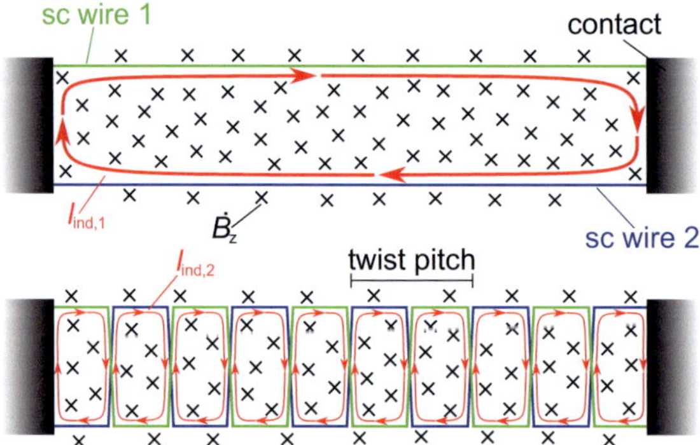

Figure 3.3: Schematic drawing of coupling currents in superconducting cables. Top: Large coupling currents ($I_{ind,1}$) in a superconducting cable consisting of two wires which are not transposed. Bottom: Low coupling currents ($I_{ind,2}$) in a superconducting cable consisting of two transposed wires.

Figure 3.4: Schematic drawing of different cable-in-conduit conductors: Roebel CICC on the left, Rutherford CICC in the center and stranded CICC with sub-cables (in red) and a central cooling channel on the right. Structural materials are shown in black, copper in brown and superconducting wires in green.

Figure 3.5: Cable in conduit conductors of the ITER magnet system: central solenoid and poloidal field coil conductor (left), toroidal field coil conductor (right). Pictures from [ENE].

3.1.3.2 Challenges in fusion magnets

In fusion experiments long containment times and large plasma volumes are desired. This results in large machines utilizing big superconducting magnets. For such containment of plasma, high magnetic fields and high precision and stability of the fields are essential. Thus superconducting fusion magnets are a challenging technology:

- Fusion magnets are among the largest magnets with diameters up to several meters.

- Magnetic fields of several tesla are needed in the center of the magnet for the plasma containment. This leads to fields on the conductor which are considerably higher.

- Generating such magnetic fields requires overall currents in the mega ampere range. This results in currents per winding of several kilo ampere.

- High currents and high magnetic fields generate huge Lorentz forces. Mechanical stabilization of fusion magnets is essential.

- A significant amount of energy is stored within the magnetic field of a fusion magnet. In the conductor, there has to be sufficient electrical stabilization to safely shut down the current after a quench.

- Depending on the coil size, the number of turns and the current shutdown rate, high discharge voltages can occur after a fault. Electrical insulation able to withstand such voltages is needed.

- A fusion magnet consists of different materials, e.g. structural materials, insulation and superconductor. By cooling such a heterogeneous structure to cryogenic temperatures mechanical stress occurs due to differences in the thermal expansion coefficients of the materials.

- Low temperature superconductors especially niobium-tin are fragile materials. They are sensitive to mechanical loads and have to be considered in the manufacturing of the magnets.

- Reliable cooling to cryogenic temperatures is required.

These challenges can and have been met in several superconducting fusion magnets. Cable-in-conduit conductor (CICC) using low temperature superconductors are able to carry kilo ampere currents in high magnetic background fields and provide sufficient cooling and mechanical stability. However, there are limits to this technology. The unfavorable mechanical properties of low temperature superconductors and their impact on the manufacturing process of the whole magnets are limiting factors in large fusion magnets. In this area there is high potential for improvement by using different superconducting materials. In the following this is shown in detail in two examples for the ITER toroidal field coils.

Heat treatment

To form the superconducting phases of A15 low temperature superconductors such as Nb_3Sn, extensive heat treatment is required. After heat treatment A15 superconductors are very brittle (see **subsection 2.2.2**). Niobium-tin (Nb_3Sn) cables are damaged by winding to coils if not located in a neutral axis. Therefore the coils of large fusion magnets (e.g. ITER) are manufactured using the "wind, react & transfer" method. In this method the complete CICC is wound to the shape of the coil, heat treated, insulated and transferred to the structural material of the coil. This means that the jacket of the CICC must also undergo heat treatment. The stainless steel (316-LN) used for the toroidal field coils, poloidal field coils and central solenoid jackets of ITER and other fusion experiments is degraded by the heat treatment. The heat treatment can lead to alloy segregation and can accumulate grain boundaries sensitizations as seen in **figure 3.6**. On the left of **figure 3.6** a compacted 316-LN stainless steel tube ITER toroidal field coil jackets is shown at 500x magnification prior to heat treatment. There is no sensitization visible at the boundaries and the maximal elongation is above 20 % which is the requirement for ITER [WJW11]. On the right of **figure 3.6**, a similar jacket is shown after compaction and heat treatment. Almost all the grain boundaries are weakened [Sgo11]. The material has become much more brittle and the maximal elongation is far below the required 20 % [WW11]. This effect depends on the composition of the melt, the associated solution annealing and the superconductor heat treatment conditions. There are melts of ITER toroidal field coil jacket material with maximal elongation below the 20 % limit and others with more than 30 % critical elongations [WJW11] after heat treatment.

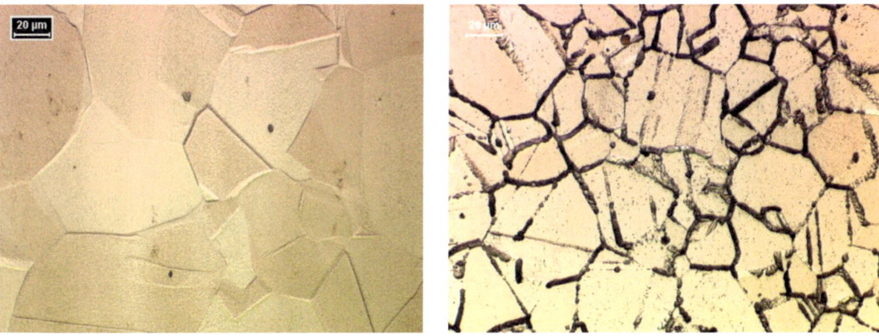

Figure 3.6: Example of 316-LN stainless steel jackets for ITER toroidal field coil cable-in-conduit conductors: no heat treatment, compacted only, as reference (left), compacted and heat treated (right). Pictures from [Sgo11].

Cyclic mechanical loading

Niobium-tin is sensitive to repeated mechanical loading, too. In the SULTAN test facility, toroidal field coil cable samples from different manufacturers exhibit a decrease in the current sharing temperature T_{cs}. Their current sharing temperatures stabilize and remain above 5.7 K (which is the T_{cs} limit of ITER) after 1200 load cycles as shown in **figure 3.7**.

Similar degradations have also been observed for the central solenoid conductor [Dev11] which is exposed to cyclical tensile loads due to the pulsed operation mode of Tokamak plasma containment systems.

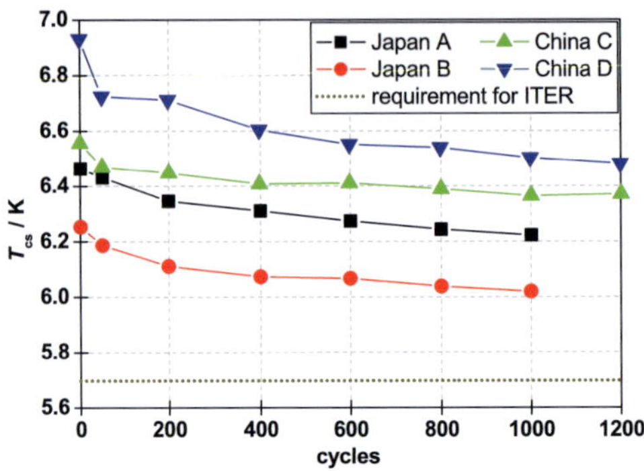

Figure 3.7: Degradation of the current sharing temperature (T_{cs}) of ITER toroidal field coil
Nb$_3$Sn conductors: samples from Japanese manufactures (referred to as A & B), samples
from Chinese manufacturers (referred to as C & D). Data from [NNY11, LWB11].

3.2 Main parameters of fusion power plants

ITER, with a planned fusion power of about 400 MW [CEA07], is at present the largest fusion
experiment [ITE10]. Its goal is to achieve a sustained controlled fusion reaction which is a
critical step in verifying that energy can be gained using controlled fusion. ITER is designed as
an experimental reactor and not as a fusion power plant. To show the usability of nuclear fusion
for electrical power generation a demonstration reactor (DEMO) is needed. This demonstration
reactor will be designed as a prototype for fusion power plants [Smi05]. It has four main goals
[Zoh10]:

- "demonstrate a workable solution for physics and technology question"

- "demonstrate large-scale net electricity production with self-sufficient fuel supply"

- "demonstrate high availability and reliability operation over a reasonable time span"

- "allow assessment of the economic prospects of a power plant"

To achieve the second goal DEMO has to deliver electrical energy in a range from which an
upscale to a commercial power plant can be done. This should be in the range of several hundred
MW, comparable to fission or coal power plants. Depending on the fraction of the recirculation
power, this means that the thermal power of such a demonstration fusion reactor must be in the
range of 2 GW to 5 GW. Different studies of how such a fusion power plant can be realized have
been conducted. Depending on the operating mode (pulsed or steady state) and the assumptions
on the plasma physics, these studies result in various sets of main parameters. For building the
magnets, the magnetic field in the plasma B_t, the major radius R_p and the minor plasma radius

a_p are important as they directly influence the maximal magnetic field on the conductor B_{max}. In the toroidal field coils the product of maximal field on the conductor B_{max} and operating current I_{op}, the Lorentz forces, are greatest. The maximal magnetic field on the conductor, B_{max}, is influenced by the distance between the plasma and the toroidal field coils (Δ_{int}). This distance is the sum of the thickness of the scrape off layer, the vacuum vessel, the blanket, the neutron shielding, the vacuum layer for the magnets and the magnet casing. The contribution of these components to Δ_{int} has been investigated for different scenarios [BC11]. To reduce the magnetic fields on the conductor, the internal distance Δ_{int} should be as low as possible. Using helium cooled blankets Δ_{int} ranges from 1.37 m to 1.45 m with an average of 1.40 m. From these parameters, the magnetic field on the conductor B_{max} can be roughly estimated using **equation 3.1** and a correction factor κ of 1.16 [HD11].

$$B_{max} = \frac{\kappa B_t}{1 - \frac{a_p}{R_p} - \frac{\Delta_{int}}{R_p}} \qquad [3.1]$$

B_t: is the magnetic field in the plasma. Increasing magnetic field in the plasma B_t also increases the maximal magnetic field on the conductor B_{max}.

R_p: is the major plasma radius. This is the radius of the center of the plasma around the torus.

a_p: is the minor plasma radius. This is the radius of the plasma in any toroidal field coil. Increasing the minor plasma radius a_p results in larger coils and in a higher the maximal magnetic field on the conductor B_{max} for a given field in the plasma.

Δ_{int}:   is the internal distance. This is the distance between the plasma and the toroidal field coils. The internal distance Δ_{int} is the sum of the thickness of blanket and shielding components, the vacuum vessel and the magnet casing. Increasing the internal distance Δ_{int} increases the size of the toroidal field coils and increases the maximal magnetic field on the conductor B_{max} for a given field in the plasma. For the magnets a low internal distance Δ_{int} is desirable.

B_{max}:   is the maximal magnetic field on the conductor. This is the field for which the superconducting cables have to designed. The maximal magnetic field on the conductor B_{max} is determined by the magnetic field in the plasma and the size of the toroidal field coils.

I_{coil}:   is the coil current. This is the overall current in a coil. It is determined by the size of the coils and the field in the plasma.

I_{op}:   is the operating current. This is the current per winding. It is determined by the coil current and the number of turns. Increasing the number of turns reduces the operating current I_{op} for a given coil current I_{coil}.

In **figure 3.8** the parameters major plasma radius R_p, major plasma radius a_p and internal distance Δ_{int}, which determine the size of the toroidal field coils, are visualized in a schematic drawing of a Tokamak magnet system in side view (on the left) and in top view (on the right).

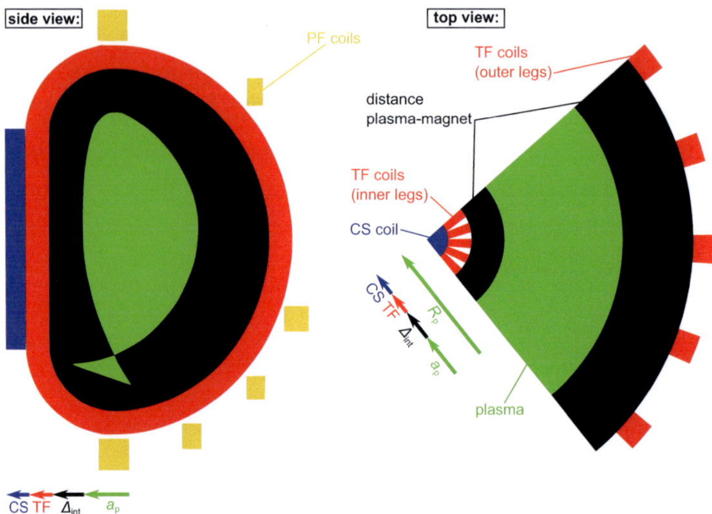

Figure 3.8: Schematic drawing of one half of a Tokamak magnet system in side view (left) and in top view (right). The central solenoid is shown in blue, the distance between the plasma (green) and the toroidal field coils (red) in black and the poloidal field coils in yellow.

For a fusion power plant, the fusion power needs to be increased by a factor of 5 and 12 compared with ITER. Various studies on how such a fusion reactor can be realized have been conducted. In **table 3.2** the main parameters of these studies are listed. The maximal magnetic field on the conductor B_{max} is essential for designing the toroidal field magnets. If in a power plant concept, the maximal magnetic field on the conductor B_{max} has not been explicitly stated, it is calculated using **equation 3.1**. In these calculations a correction factor κ of 1.16 and an average internal distance Δ_{int} of 1.4 m are assumed, if not given otherwise in the corresponding study.

These sets of main parameters of fusion power plants have to be compared with existing fusion experiments. In **figure 3.9** the maximal magnetic field on the conductor of the toroidal field coils and the minor plasma radius are shown for all existing superconducting fusion experiments and the DEMO concepts.

In these studies of fusion power plants the increase in fusion power comes with an increase in reactor size compared with ITER. Most concepts use a major radius of 7.5 m (ITER: 6.2 m) and a larger minor plasma radii of more than 2 m (ITER: 2 m). In addition, the magnetic field in the plasma is increased compared with ITER resulting, in all concepts, in higher maximal magnetic field on the conductor B_{max}, ranging from 12.3 T to 18.2 T.

Table 3.2: Parameters of different DEMO concepts. Only the concepts which allow the determination of the maximal magnetic field on the conductor are listed. Values in black are from references. Values in red are averaged values. Values in blue are measured from dimensional drawings. Values in green are calculated.

study	R_p / m	a_p / m	Δ_{int} / m	B_T / T	B_{max} / T	references
ITER	6.2	2.0		5.3	11.8	[HD11]
DEMO (2006)	7.5	2.46	1.4	5.86	14.4	[DH08, Duc]
DEMO1 (2011)	10.0	2.5	1.8	7.0	12.3	[Zoh11, Boc11]
DEMO2 (2011)	8.5	2.84	1.8	5.75	14.7	[Zoh11, Boc11]
DEMO1 (2012)	9.0	2.25		7.06	13.45	[Kem12a, Kem12b]
Model AB (inductive)	9.55	3.15	1.4	7.0	15.5	[Gir06]
Model AB (advanced)	7.5	3.0	1.4	6.0	16.8	[Gir06]
PPCS "Model A"	9.55	3.2		7.0	13.1	[MCP06, MCP05]
PPCS "Model B"	8.6	2.9		6.9	13.2	[MCP06, MCP05]
PPCS "Model C"	7.5	2.5	1.45	5.6	13.6	[MCP06, MCP05, BC11]
PPCS "Model D"	6.1	2.0		5.6	13.4	[MCP06, MCP05]
Garcia	7.5	2.5	1.4	6.0	14.5	[GGA08]
Pacher	8.1	2.8	1.4	5.7	13.7	[PPJ07]
Kolbasov DEMO-S (basic)	7.8	1.5		7.7	14.3	[KBB08, SKS00]
Kolbasov DEMO-S (adv.)	1.5	1.5		8.75	16.0	[KBB08, SKS00]
Sokolov DEMO-S	9.0	2.0	1.4	9.0	16.8	[Sok94]
Sokolov DEMO-P	6.95	2.1	1.4	6.86	16.0	[Sok94]
Tobita "slim-CS"	5.5	2.1		6.0	16.4	[TNT09, TNE06]
DEMO-Crest	7.25	2.13		7.8	16.0	[HOA05, HOAO07, IMH10]
ARIES-AT	6.86	1.3	1.26	5.2	13.4	[WNT06, NAB06]
PPCS-C	7.5	2.5	1.4	6.4	15.5	[CKW11]
HCSB "OP1"	9.0	2.0	1.4	8.23	15.3	[FZZ09]
HCSB "OP2"	8.0	2.2	1.4	7.31	15.4	[FZZ09]
HCSB "OP3"	7.2	2.1	1.4	6.86	15.1	[FZZ09, GKG09]
SSTR	7.0	1.5			16.5	[KSO91]
JAERI A "CS-less"	5.1	2.1		5.6	18.2	[TNE06]
JAERI B "slim CS"	5.5	2.1		6.0	16.4	[TNE06, TNE09]
JAERI "full CS"	6.5	2.1		6.8	14.6	[TNE06]
Indian DEMO	7.7	2.6	1.4	6.0	14.5	[SD08]

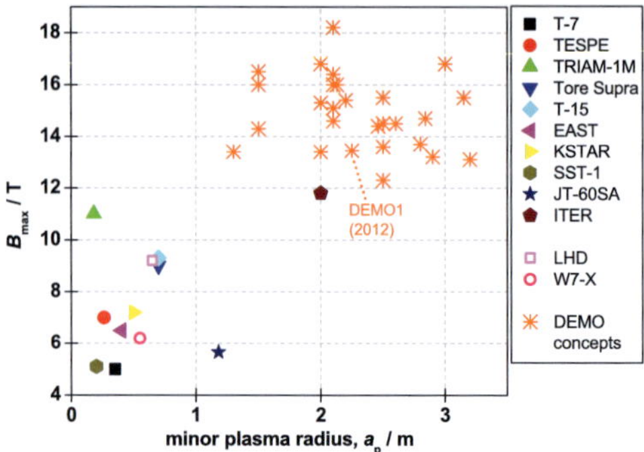

Figure 3.9: Overview of existing superconducting Tokamak type (closed symbols), Stellarator type (open symbols) fusion experiments and concepts for fusion power plants. Maximal magnetic field on the toroidal field coil conductor (B_{max}) vs. minor plasma radius (a_p).

3.3 Conductors in fusion power plants

Building magnets of the size relevant for fusion power plants is challenging. The electrical and mechanical requirements for withstanding the enormous Lorentz forces and for allowing safe shutdown in the case of a quench are discussed in **subsection 3.3.1**. The options for cooling the conductors are assessed in **subsection 3.3.2**. In fusion magnets, there are high neutron fluxes on the conductor, nuclear activation is investigated in **subsection 3.3.3**.

3.3.1 Electrical and mechanical requirements

The electrical and mechanical requirements of conductors increases with the size and the field of the magnets. Main parameters of the coils of superconducting fusion experiments are compared in **figure 3.10.** The current per winding vs. the maximal magnetic field on the conductor is shown on the left of the figure and the Lorentz forces vs. the (planned) year of completion on the right.

With an operating current I_{op} of 68 kA and a maximal magnetic field on the conductor B_{max} of 11.8 T, the current, Lorentz forces (802 kN m^{-1}) and stored energy (2.28 GJ per coil) are highest in the toroidal field coils of ITER [ITE09b]. Fusion power plants are discussed on the basis of the ITER magnet system, for which the main parameters are given in **table 3.3**.

The Lorentz forces per conductor length F_L are proportional to the operating current I_{op} and the maximal magnetic field on the conductor B_{max}, thus, for a given size and field they depend on the coil current I_{coil} and the number of turns n. The relation between operating current I_{op}, coil current I_{coil} and number of turns n_{TF} is given in **equation 3.2**.

$$I_{op} = \frac{I_{coil}}{n_{TF}} \qquad [3.2]$$

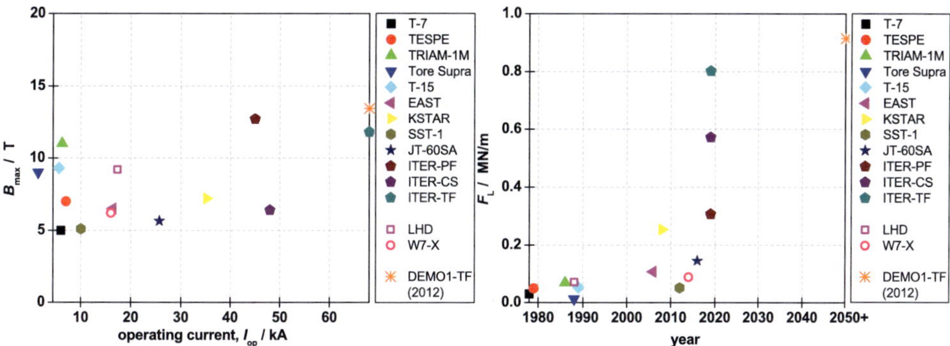

Figure 3.10: Comparison of superconducting fusion experiments and the concept "DEMO1" from 2012. Maximal magnetic field on the conductor (B_{max}) vs. current per winding of the toroidal field coils (I_{TF}) on the left. Lorentz forces (F_{L}) vs. (planned) year of completion on the right.

Table 3.3: Design parameters of the ITER magnet system. Data from [ITE09b, ITE09c, ITE09d].

	I_{c} / kA	B_{max} / T	T_{op} / K	$F_{\mathrm{L,\,max}}$ / kN m^{-1}	energy per coil / GJ	number of strands
CS	40 - 45	13.0 - 12.7	4.5	520 - 572	1 - 1.15	576 Nb$_3$Sn, 288 Cu
PF 1 - 6	50 - 48	5.0 - 6.4	4.5	250 - 307	0.9 - 1.3	1440 - 720 NbTi, Cu core
TF	68	11.8	5.0	802	2.28	900 Nb$_3$Sn, 522 Cu

The operating current is a design parameter as the magnetic field can either be produced with a high operating current in a few turns or with low operating current in many turns. Lower operating current reduces the Lorentz forces per conductor length but increases the inductivity of the magnet. Higher inductivity increases the discharge voltages and discharge times of the magnet, requiring better electrical insulation and more electrical stabilization to be operated safely. The additional stabilization increases the non superconductor fraction in the cable and decreases the engineering current density J_{e} of the winding pack.

The cross section areas and the mechanical properties of the components of the winding pack of the inner legs of the toroidal field coils of ITER are shown in **table 3.4**.

The mechanical stabilization, radial plates and jacket of the toroidal field conductor require about 57.1 % of the total cross section area of the winding pack. The remaining area is shared by the superconductor (Nb$_3$Sn : 16.2 %), the electrical stabilization (Cu: 14.6 %), void area (cooling channel and void fraction of cable: 12.2 %) and insulation (not considered in the following). The yield strength of the structural materials is with more than 900 MPa significantly higher than of the superconductor (80 - 100 MPa) and the electrical stabilization ($\approx$ 34 MPa). Therefore, neither the electrical stabilization nor the superconductor contribute to the overall mechanical stabilization. The yield strength of the total winding pack is reduced to 535 MPa.

To distribute the load, each winding must be individually mechanically stabilized with structural

Table 3.4: Components of the inner legs of the toroidal field coils of ITER. The cross section areas are per turn (134 turns). Electrical insulation is not considered due to the negligible cross section area. Data from [ITE09b, ITE09e, Wei07][Jos99, p. 84].

	cross section area	fraction	yield strength
Nb_3Sn	$567.4\,mm^2$	16.2 %	80 - 100 MPa
stabilization	$508.2\,mm^2$	14.6 %	≈ 34 MPa
structure - radial plates	$1722.5\,mm^2$	49.3 %	> 900 MPa
structure - jacket	$270\,mm^2$	7.7 %	> 950 MPa
void area	$424.3\,mm^2$	12.2 %	-
total	$3492.5\,mm^2$	100 %	> 535 MPa

material to avoid an outbound accumulation of forces and to prevent the inner windings from being crushed by the outer ones. Even inside the cable-in-conduit-conductor, the forces are concentrated and distributed inhomogeneously. Strands located to the inside of the magnet experience compressive stresses, while strands on the outside experience tensile stresses. In the toroidal field coil conductors of ITER a degradation of the carrying capabilities has been observed after cycling loading. Damage is observed in areas with high tensile stresses, i.e. the outer strands [LWB11].

In an HTS fusion magnet, there is a significant amount (50 % of cross section area in Super-Power tapes) of structural material in *REBCO* tapes. The Hastelloy® C-276 substrate of these tapes has a yield strength of at least 1.2 GPa at 4.2 K [Ohs08]. This is above the requirements for structural material of ITER (yield strength at 4.2 K of more than 900 MPa) and therefore fully qualifies as structural material. It is in direct contact to the superconductor and can distribute mechanical loads to avoid stress concentrations. *REBCO* tapes for magnet or cable applications, also consist to a large fraction of copper (40 % copper and 4 % silver cross section area in stabilized SuperPower tapes), which contributes to the overall electrical stabilization of the cable.

This is summarized for the example of the fusion power plant concept DEMO1 from 2012. In that concept the maximal magnetic field on the conductor is increased by 14 % to 13.45 T, while keeping the same operating current as ITER of 68 kA. This results in an increase of the number of turns n_{TF} by the same factor to 153.

- The Lorentz forces per meter cable are increased by 14 % to 914.6 kN m^{-1}. To compensate the forces, the amount of structural material per winding has to be increase by 14 %, as well. This leads to a total increase of the structural material in the winding pack of 30 %.

- The higher number of turns and the larger diameter magnets result in a higher inductance of the coils and in more stored energy. Either the electrical insulation has to be improved or the discharge times increased. With the same insulation, at least 14 % higher discharge times and amount of electrical stabilization per winding are necessary. In the winding pack, this results in an 30 % increase of the total amount of electrical stabilization.

Both changes increase the non-superconductor area and thus reduce the engineering current density J_e in the winding pack. For the engineering current density J_e to remain close to ITER (ITER: 11 A mm^{-2} [HD11]) an increase of the current density of the superconductor is essential. With the superconductor technology of today, higher engineering current densities J_e in the winding packs than ITER, as assumed in some fusion power plant concepts (fusion power plant concept CREST: $J_e = 15$ A mm^{-2} [HOAO07]), are unlikely.

3.3.2 Cooling options

For stable operation of a fusion magnet an efficient cooling of the superconducting coils is essential. In the magnets of a fusion reactor there are different heat sources. There is the conductive heat from the casing, the heat deposited with the neutron flux, the resistive loss in the conductor joints and an AC loss in the superconductor due to plasma disruptions and the pulsed operation mode of the central solenoid. These heat inputs have to be compensated by the cooling system of the coils. As shown schematically in **figure 3.11**, three different cooling types are possible: pool cooling (left), forced-flow cooling (center) and indirect cooling (right).

- The helical coils of the Large Helical Device (LHD) are pool-cooled with liquid helium [MSI06b]. In the large coils of a fusion power plant pool-cooling is not applicable and will not be considered.

- Forced-flow cooled conductors have a higher rigidity and dielectric strength than pool-cooled conductors. Therefore most modern fusion magnets including the Poloidal coils of the Large Helical Device [TMC03], Wendelstein 7-X [Sap00], SST-1 [Sax00], JT-60SA [IBK10] and ITER [ITE] use forced-flow cooled coils. For forced-flow cooling a cryogenic cooling agent has to be compressed and circulated. A forced convective flow of the cooling agent through the conductor is provided. Superconducting cables require voids or cooling channels for forced-flow cooling. In a large coil, the pressure drop occurring in the cooling channels [TIM00] limits the unit conductor length, and a large number of pipes and conductor joints must be installed. Forced-flow cooling is quite complex but enables stable temperatures with low deviations.

- Indirect cooling is a promising candidate for future fusion magnets because it solves the issue of pressure drops [MSI06a]. In indirect cooling, the superconductor is cooled from the outside of the superconducting cable through heat conduction. In general, indirect cooling is simpler, but as the heat has to be conducted through the jackets of the superconducting cable-in-conduit conductors, the temperature margins are higher

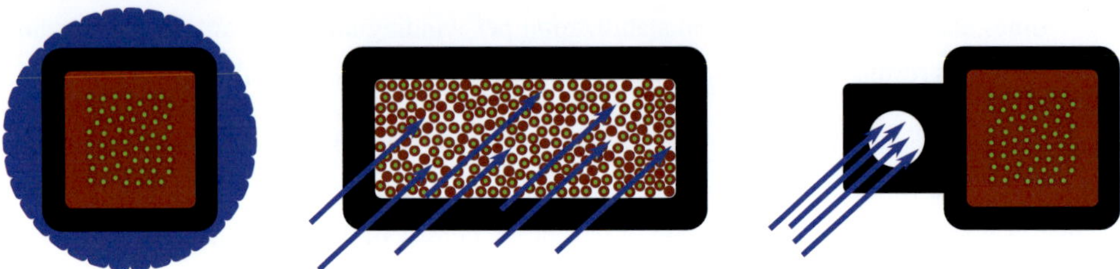

Figure 3.11: Schematic drawing of different cooling types of superconducting magnets: Pool cooling on the left, force-flow cooling in the center and indirect cooling on the right. Structural materials are shown in black, copper in brown, superconducting wires in green and cooling agents in blue.

compared with forced-flow cooling. At present, no indirect cooled coils have been constructed for fusion experiments, even though indirect cooling is commonly used in accelerator magnets [Yam04].

Cooling solutions are mainly determined by the application, the size of the magnet and the superconductor material. The temperature margin of the superconductor plays a major role, especially between forced-flow and conduction cooling. As shown in **figure 3.12**, there are high differences in the temperature dependence of the engineering current density J_e for Nb_3Sn and REBCO tapes.

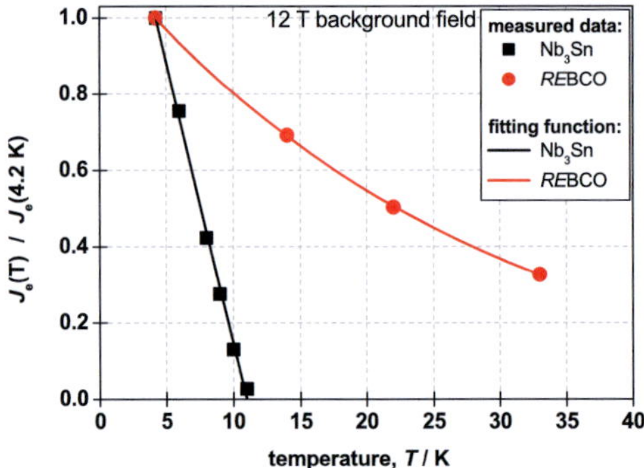

Figure 3.12: Measured (points) and fitted (lines) temperature dependence of the normalized engineering current density J_e of Nb_3Sn and REBCO at magnetic background fields of 12 T. Data from [GKK01, Haz12].

The measured temperature dependence is fitted in **equation 3.3** and **equation 3.4**. Fitted data and measured data are in good agreement for both materials in the whole temperature range.

$$J_e^{Nb_3Sn}\,(T, 12\,T) = -0.148 \cdot (T - 10.99) \cdot J_e^{Nb_3Sn}\,(4.2\,K, 12\,T) \qquad [3.3]$$

$$J_e^{REBCO}\,(T, 12\,T) = \left(1.22 \cdot \exp\left(\frac{-T}{27.81}\right) - 0.05\right) \cdot J_e^{REBCO}\,(4.2\,K, 12\,T) \qquad [3.4]$$

The functions can be used to compare the influence of a temperature variation around 4.2 K and to calculate temperature margins ΔT for defined minimal normalized current carrying capabilities $J_e(T)/J_e(4.2\,\mathrm{K})$. Temperature margins ΔT of Nb$_3$Sn and *REBCO* are given in **table 3.5** at 95 %, 90 %, 85 %, 80 %, 75 % and 50 % of the engineering current densities at 4.2 K.

The temperature margin of *REBCO* is at least four times that of Nb$_3$Sn. Because of this, magnets utilizing *REBCO* tapes are much better suited for conduction cooling. The larger temperature margin of conduction cooling has less impact on the minimal engineering current density.

Furthermore, cooling agents other than helium are possible in *REBCO* magnets. Boiling points, molar heat capacities and heat of vaporization are given for the cryogenic cooling agents helium, hydrogen, neon and nitrogen in **table 3.6**.

Increasing the operating temperature improves the energy efficiency of cooling processes. An ideal inverse Carnot cycle requires 70.7 W of mechanical work at 300 K to remove 1 W of heat at 4.2 K. At 77 K, the required mechanical work is reduced to 3.3 W [Riz12]. In applications, this difference is even larger due to the less than ideal effectiveness of cryogenic plants. For example, operating ITER-like magnets at 50 K instead would reduce the electrical cooling power from 19 MW to 12 MW [Hel07, p. 49]. However, higher operating temperatures e.g. reduce the yield strength of the structural materials, so more structural materials are necessary for the same level of mechanical stabilization. This has to be considered carefully.

Table 3.5: Temperature margins of state-of-the-art Nb$_3$Sn and *REBCO* at 95 %, 90 %, 85 %, 80 %, 75 % and 50 % normalized engineering current densities J_e at 12 T magnetic background field. Data calculated with **equation 3.3** and **equation 3.4**.

fraction of $J_e(4.2\,\mathrm{K})$	95 %	90 %	85 %	80 %	75 %	50 %
temperature margin of Nb$_3$Sn	0.3 K	0.7 K	1.0 K	1.3 K	1.7 K	3.4 K
temperature margin of *REBCO*	1.3 K	2.6 K	4.2 K	5.8 K	7.6 K	18.1 K

Table 3.6: Boiling points, molar heat capacities and heat of vaporization of the cryogenic cooling agents helium (He), hydrogen (H), neon (Ne) and nitrogen (N). Data from Wikipedia, the free encyclopedia.

	He	H	Ne	N
boiling point / K	4.22	20.28	27.07	77.36
molar heat capacity / J mol^{-1} K	20.79	28.84	20.79	29.12
heat of vaporization / kJ mol^{-1}	0.08	0.90	1.71	5.56

3.3.3 Nuclear activation

In fusion experiments and fusion power plants there are high neutron fluxes. The nuclear activation of the components of the reactor has to be considered as neutron fluxes influence the material properties. There are three major aspects of nuclear activation:

- In structural materials and the blanket it can lead to outgassing of helium and can degrade the mechanical properties. Therefore reduced-activation-ferritic-martensitic steels (e.g. EUROFER) [LMS02] are mainly used.

- In the coil, the neutron flux influences the current carrying capabilities of the superconductors [UIM09]. The current carrying capabilities of *RE*BCO tapes are improved through neutron irradiation. The radiation causes defects which act as additional flux pinning centers. These increase the pinning forces and have positive effect on the critical current density in magnetic background fields [AUI11].

- Additionally, long term activation renders the materials radioactive. For a fusion power plant it is desired that all activated materials can be safely recycled after 100 years of cooling [Hel07, p. 29] to allow decommissioning of the reactor without creating nuclear waste which has to be stored for millenia. This is determined by the decay of the contact dose rates, which is strongly material dependent.

The decay of contact dose rates is summarized in the following for the constituent materials of the low temperature superconductor material Nb_3Sn and the high temperature superconductors *RE*BCO tapes and BSCCO tapes and round wires. Data from [Hel07] is used.

Decay of the contact dose rates of the constituents of Nb_3Sn
The composition of niobium-tin superconductor wires depends on the manufacturing process, in internal tin and powder-in-tube processes the superconductor filaments mainly consists of niobium (Nb) and tin (Sn). In bronze route filaments, the constituent materials are niobium (Nb) and bronze, which is an alloy of copper (Cu) and tin (Sn). The filaments are usually embedded in a copper (Cu) matrix. The decay of the contact dose rates of these materials, after five years of nuclear activation with the first wall dose (FWD) of ITER, are shown in **figure 3.13**.

The contact dose rates of tin and copper are at the recycling limit after 100 years, the decay of nuclear activity in niobium is much slower. Its contact dose rates remain above the recycling limit for more than one million years. Niobium has to be avoided in fusion magnets if possible.

Decay of the contact dose rates of constituents of *RE*BCO tapes
The largest fraction of *RE*BCO tapes are the substrate (Hastelloy or Ni-alloy) and the copper stabilization (if existing). The other layers are usually insignificantly thin (few μm range). The decay of the contact dose rates after five years of nuclear activation with the first wall dose (FWD) of ITER is shown in **figure 3.14** for the constituent materials of *RE*BCO tapes.

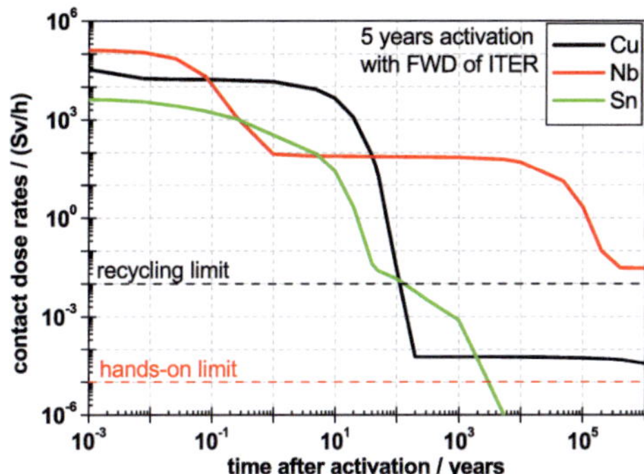

Figure 3.13: Contact dose rates of the elements found in Nb_3Sn after 5 years of nuclear activation with the first wall dose (FWD) of ITER. Data from [Hel07].

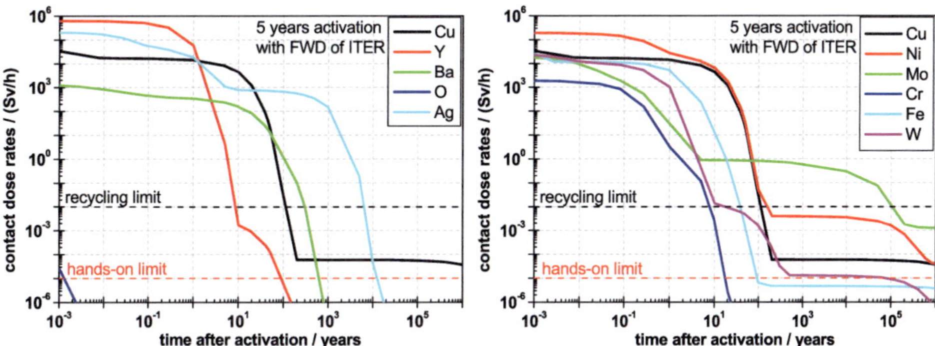

Figure 3.14: Contact dose rates of the elements found in *RE*BCO after 5 years of nuclear activation with the first wall dose (FWD) of ITER. Data from [Hel07].

There are tapes with substrates of nickel alloys (e.g. nickel-wolfram) or Hastelloy. Hastelloy mainly contains nickel (Ni), molybdenum (Mo), chromium (Cr), iron (Fe) and wolfram (W). The contact dose rates of these materials are at or below the recycling limit except for molybdenum. In fusion magnets *RE*BCO tapes with a low fraction of molybdenum are preferable.

The constituent materials of the superconducting layer are also below the recycling limit. Silver (Ag) however remains above the recycling limit for more than one thousand years at this level of nuclear activation. In *RE*BCO tapes from most manufacturers, the silver layers are very thin (SuperPower, Bruker ETS and AMSC: 1 - 3 µm range) and the silver content is negligible.

Decay of the contact dose rates of constituents of BSCCO tapes and wires

A large fraction (usually more than 50 %) of BSCCO 2212 round wires and BSCCO 2232 tapes is the silver, silver-gold or silver-magnesium matrix. Silver (Ag) remains nuclear active for more than one thousand years (see **figure 3.14**), making BSCCO round wires and tapes unsuitable conductors in high neutron flux environments such as fusion magnets.

3.4 Potential of high temperature superconductors in fusion magnets

High temperature superconductors, especially *RE*BCO tapes (see **subsection 2.3.3**) are materials with impressive current carrying capabilities, in field performance and mechanical stability. With these materials it is possible to meet the challenges of large fusion magnets. *RE*BCO has the potential to reach higher fields, to reduce the amount of structural materials, to prevent degradation of the conductor, avoid "wind & react" or "wind, react & transfer" manufacturing, to use cooling agents other than helium, to avoid forced-flow cooling, to increase the cooling efficiency and to simplify the cryostat required.

- In *RE*BCO the critical current density is far less influenced by increasing magnetic background fields in comparison with Nb_3Sn and NbTi. The $J_c - B$ dependence of *RE*BCO is shallower than that of Nb_3Sn or NbTi as shown in **figure 2.7**. This is essential for fusion magnets magnets with maximal magnetic fields higher than ITER ($B_{max} > 11.8\,T$). In high magnetic background fields *RE*BCO can provide higher current densities than Nb_3Sn or NbTi. This is essential in fusion power plants as fusion power plants are commonly assumed to have larger magnets and higher fields than ITER (see **figure 3.9**).

- In Tokamak plasma confinement systems the product of magnetic field on the conductor and current is highest in the toroidal field coils. Thus mechanical stabilization is critical in these coils. With about 550 - 700 MPa [HRB11] the critical stress limit of *RE*BCO is up to 7 times higher than that of Nb_3Sn (80 - 100 MPa [KKS01, SFAF07, Wei07] range as shown in **figure 2.10**). Additionally, the critical strain limit (less than 5 % degradation of the current carrying capabilities due to mechanical tensile strain) of *RE*BCO is with 0.5 % much higher than that of Nb_3Sn. Due to these significant advantages of the mechanical properties, the fraction of additional structural materials is expected to be lower in magnets using *RE*BCO compared with Nb_3Sn magnets. This allows for either the construction of more compact magnets or magnets with higher fields.

- If cables with multiple superconductor strands or tapes are used in magnets, their mechanical loads are not distributed homogeneously. Lorentz forces of the inner strands (facing the plasma) or tapes are transferred to the outer ones (facing away from the plasma). Therefore mechanical loads are accumulated towards the center of a Tokamak magnet system which can lead to dangerous stress concentrations. In magnets using *RE*BCO each tape is provided with a degree of mechanical stabilization directly at the conductor due to the substrate of *RE*BCO (see **figure 2.13**). Thus, the accumulation of mechanical loads is less severe in *RE*BCO magnets.

- Magnets using Nb_3Sn are commonly assembled using "wind & react" or "wind, react & transfer" due to the brittle nature and unfavorable mechanical properties of fully formed Nb_3Sn [MS08]. For large fusion magnets this is especially challenging as the whole coil has to undergo heat treatment. This requires huge ovens and can degrade the structural materials of the coils [RBBo10] (see **subsection 3.1.3.2**: heat treatment). *RE*BCO tapes are mechanically strong materials (critical stress: $> 550\,\mathrm{MPa}$ [Supd]) which are flexible and can be bent to low radii (critical bending radius: $> 11\,\mathrm{mm}$ [Supd]). *RE*BCO is fully processed prior to magnet construction. Therefore magnets utilizing *RE*BCO can be assembled using "react & wind" techniques [TBL07, TJH10]. This drastically simplifies the construction of *RE*BCO magnets compared with magnets made of Nb_3Sn.

- Critical temperatures of *RE*BCO are in the 90 - 100 K [WAT87, TA09] range depending on the composition. In HTS magnets, electrical cold tests are possible with nitrogen cooling, allowing the evaluation of the performance of the conductor, the electrical resistivity of the joints, the thermal transfer and the operation of the safety system. These tests are possible even if the nominal operating temperature of an HTS magnet is far below the boiling point of nitrogen by extrapolating the conductor performance. In an LTS magnet, electrical cold test have to be done with liquid helium [LCD05] requiring a complex test setup and being expensive [MMH09].

- Helium is scarce on earth [DPG11]. The natural helium reserves on earth are expected to be depleted within 30 years [Mac11]. For fusion power plants it is therefore beneficial that cooling agents other than helium can be used in the magnets. This is not possible in LTS magnets due to the low critical temperatures of niobium-titanium (NbTi: 9.1 - 9.5 K [FHKa]) and niobium-tin (Nb_3Sn : up to 18.3 K [FHKb]), in contrast to *RE*BCO which is used for technical applications with temperatures up to 77 K. This enables different cooling agents as neon (boiling point: 27.07 K[a]), hydrogen (boiling point: 20.28 K[b]) or nitrogen (boiling point: 77.36 K[c]) to be used.

- Energy efficiency in cooling processes strongly depends on the temperature difference. In an ideal inverse Carnot cycle, 70.4 W mechanical work at 300 K are required to remove 1 W of heat at 4.2 K. At 30 K, the mechanical work necessary for 1 W of cooling power is reduced to 8.1 W and at 50 K to a mechanical work of 5.0 W [War77, p. 764]. In cryogenic plants the cooling efficiency is even lower due to less than ideal effectiveness of the machines [Tan06, p. 84]. For cryogenic systems this means that the electrical power required for cooling depends on the amount of heat which has to be removed and the

---

[a]**Neon** from Wikipedia, the free encyclopedia
[b]**Hydrogen** from Wikipedia, the free encyclopedia
[c]**Nitrogen** from Wikipedia, the free encyclopedia

operating temperature. Fusion magnets utilizing *RE*BCO can be operated at temperatures at which low temperature superconductors are already in normal conducting state, enabling operation of HTS magnets at increased cooling efficiency.

- Currently most superconducting fusion magnets use forced-flow cooling (e.g. Wendelstein 7-X [Sap00], SST-1 [Sax00], JT-60SA [IBK10] and ITER [ITE]). In forced-flow cooled systems the pressure drop in the cooling channels [TIM00] limits the unit conductor length. Large forced-flow cooled magnets are therefore complex, requiring a high number of pipes and conductor joints. Due to the higher critical temperatures of *RE*BCO tapes, larger temperature margins are possible. Thus enabling alternative, less complex, cooling methods such as conduction cooling.

- Cryostats of low temperature superconductor magnets are in general equipped with radiation shields cooled with liquid nitrogen. A radiation shield reduces the amount of heat transported to the inside of the cryostat. Vacuum shields are needed between the nitrogen cooled radiation shield (about 77 K) and the helium cooled LTS magnet (about 1.8 - 5 K) to minimize the thermal transport. This results in a complex cryostat. A *RE*BCO magnet would not require a nitrogen cooled radiation shield if operated close to the boiling point of liquid nitrogen (e.g. 50 - 77 K). A single vacuum insulated layer is sufficient in such a case, reducing the number of shielding layers from 3 to 1. This results in less complex cryostats.

*RE*BCO tapes are promising conductors for future fusion magnets. The superior properties of *RE*BCO are essential in reaching the fields, the efficiency and reliability of the magnets required for fusion power plants. An in-depth investigation of *RE*BCO over a large temperature, mechanical strain and magnetic field range is needed to determine suitable operation scenarios in magnets. Due to the flat tape shape, new cabling concepts are required as it is not possible to assemble stranded cables directly from tapes.

4 Materials of high temperature superconductor cables

High temperature superconductor cables are heterogeneous structures consisting of various kinds of materials: superconductors, materials to provide mechanical stabilization, materials to insulate the cable or tapes and materials to fill voids in the cable. Due to the cryogenic operating temperatures of HTS cables, the thermal properties of all materials are important. It has to be be possible to cool the cable from room temperature to operating conditions (4.2 - 77 K) without damaging components. Large mismatches of the thermal expansions of adjacent materials have to be avoided. Especially sensitive are the superconducting *REBCO* tapes. Additionally, a stable operating temperature is needed, requiring an appropriate cooling system. For designing a cooling system of a superconductor cable, the thermal conductivities of all materials incorporated into the cable have to be considered.

Thermal properties are investigated in the following, thermal expansion in **section 4.1** and **section 4.2** and thermal conductivity in **section 4.3**.

4.1 Thermal expansion

In solid state matter all atoms oscillate around their center of gravity. The frequency and amplitude of these oscillations depends on the temperature and with increasing temperatures the oscillations increase. The oscillations are guided by the atomic potential. This potential is, due to the very steep repellant part in the case of the convergence of atoms, not symmetric. Thus the oscillations are not harmonic. At increased temperatures, which result in larger amplitudes of the oscillations, the average distances between atoms increase. For most solid state matter, an increase in temperature therefore results in an increase in volume and a decrease in density. This effect is called thermal expansion. The magnitude of the thermal expansion can strongly differ between different kinds of solid state materials. For a given temperature difference $(T - T_0)$, the change in length of a material $(L(T) - L_0)$ divided by its length at room temperature L_0 is the thermal expansion $\varepsilon_{TE}(T - T_0)$ of that material (see **equation 4.1**).

$$\varepsilon_{\mathrm{TE}}(T - T_0) = \frac{L(T) - L_0}{L_0} \qquad [4.1]$$

In HTS cables, temperature shifts from room temperature to operating temperatures (4.2 - 77 K) are unavoidable resulting in temperature differences ΔT of 200 - 289 K. Therefore, low thermal expansion mismatches between all materials are necessary, otherwise thermal strains and stresses can damage the cable. HTS cables consist of different of material types such as superconductor tapes, electrical stabilization, mechanical stabilization, insulation, fillers and solder, requiring an in-depth investigation of their thermal properties.

4.1.1 Experimental setup

Thermal expansion measurements are performed in a bath cryostat which is equipped with several horizontal shields to minimize the convection. The sample and the measurement equipment are located on the bottom of the cryostat. The cryostat is cooled down fast with liquid helium in about 15 - 30 min. Liquid helium is filled in the cryostat until the sample is completely submerged. After cool down the helium flow is deactivated. The cryostat now warms up again very slowly in about 12 - 24 h. Due to the slow warm up rate, the sample and the bottom part of the cryostat are in thermal equilibrium.

Two class 1 "CryoMaK"[a] extensometers are attached to the sample. During warm up the extensometers constantly determine the length of the measurement section of the sample. Their signal is amplified and registered using a high precision 24 bit voltage meter. The sample temperature is measured continuously with a calibrated DT-670-SD temperature sensor diode[b] [Lakb] from LakeShore. For protection, the temperature sensor is installed in a brass enclosure. The temperature measurement is done in a four point configuration with separate wires for the driving current and the sensor signal. A model M-218 temperature monitor from LakeShore is used to drive and read out the sensor. With this setup, a temperature resolution of at least 10.5 mK and a temperature accuracy of ± 159 mK are typically achieved [Lakc].

CryoMaK extensometers consist of a flat U-shaped copper-beryllium bar. Both ends have sharp points to prevent sliding of the extensometer on the sample. The base of the U-shape is thicker than the sides, thus a change in length of the sample results in bending of the sides. This bending is registered with 350 Ω strain gauges. Four strain gauges are used for each extensometer. They are connected in a full bridge circuit to improve the accuracy of the reading. As the sides of the extensometer are bent the resistivity of the strain gauges changes. The full bridge of the strain gauges is connected in a four point measurement setup to a high precision, low noise amplifier. Separate wires are used for the driving current and the voltage readout. These extensometers are reliable, can be used at cryogenic temperatures and deliver a resolution of at least 20 nm [Nyi05]. This setup [BWWW12] is shown schematically in **figure 4.1**.

---

[a]CryoMaK is an abbreviation for "Cryogenic Materialtests Karlsruhe". The CryoMaK lab is part of the Institute for Technical Physics (ITEP) of the Karlsruhe Institute of Technology (KIT).
[b]1.4H calibration from LakeShore

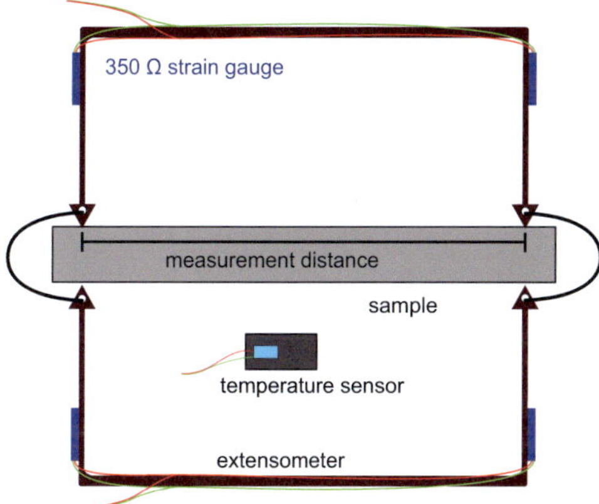

Figure 4.1: Setup of thermal expansion measurements. Two class 1 "CryoMaK" extensometers are attached to a sample. The extensometers constantly determine the length of the sample while a DT-670 temperature sensor measures the temperature of the system.

As the measurement is done in thermal equilibrium of the sample, the extensometers and the bottom of the cryostat, the extensometers also change their temperature from 4.2 K to room temperature during measurement. Thus, the signal gained from the extensometers has to be corrected to compensate for their temperature dependence. The temperature dependence of an extensometer consists of two parts.

- The calibration factor $a(T)$, which determines the distance change per voltage change, is temperature dependent.

- The material of which the extensometers are made, e.g. copper-beryllium or a titanium alloy, has a certain thermal expansion.

To compensate the temperature dependence of the calibration factor, CryoMaK extensometers are calibrated at room temperature by measuring the change in readout voltage for defined changes in distance. A calibration factor at room temperature a_{RT} is gained. To determine the temperature dependence, the extensometers are installed in a tensile machine with a temperature variable cryostat. The tensile machine repeatedly determines the calibration factor at different temperatures. By normalizing these temperature dependent calibration factors $a_n(T_n)$ with the calibration factor at room temperature a_{RT}, the correction factor $\kappa_n(T_n)$ is obtained for the temperatures T_n as shown in **equation 4.2**.

$$\kappa_n(T_n) = \frac{a_n(T_n)}{a_{RT}} \qquad [4.2]$$

A polynomial function is used to fit the data. This function corresponds to the temperature dependence of the calibration and is the temperature correction factor $\kappa(T)$ of an extensometer.

The data and the second order polynomial fit for the extensometers which are used in the thermal expansion measurement facility are shown in **figure 4.2**.

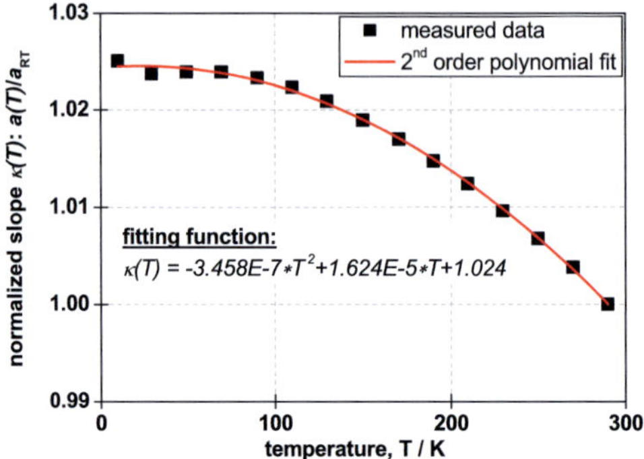

Figure 4.2: Temperature correction factor of the extensometers used in the thermal expansion measurement facility. Data from [Kel08].

The temperature correction factors can differ between different extensometers. For the extensometers used in the thermal expansion measurement facility the fitting function of **equation 4.3** is used.

$$\kappa(T) = -3.458 \cdot 10^{-7} \cdot T^2 + 1.624 \cdot 10^{-5} \cdot T + 1.024 \qquad [4.3]$$

This correction factor can now be used to determine the sample length change as shown in **equation 4.4**.

$$\Delta L_{\text{TE}} = \Delta U_{\text{ext}} \cdot a(T) = (U_{\text{ext}}(T) - U_{\text{ext,RT}}) \cdot \kappa(T) \cdot a_{\text{RT}} \qquad [4.4]$$

A change in length due to different temperatures ΔL_{TE} divided by length at room temperature L_0 is the thermal expansion $\varepsilon_{TE}(T)$ which is usually given in in $\mu\text{m m}^{-1}$ or in percent (**equation 4.5**).

$$\varepsilon_{\text{TE}} = \frac{\Delta L_{\text{TE}}}{L_0} \qquad [4.5]$$

To determine the thermal expansion of a sample, the thermal expansion of the extensometers themselves has to be corrected. This is done with a baseline measurement of a material with an extremely low thermal expansion. Zerodur^{TMc} with an thermal expansion coefficient of $0 \pm 1 \cdot 10^{-7}$ K^{-1} is commonly used. The baseline measurement is subtracted from all other thermal expansion measurements to correct for the thermal expansion of the extensometers themselves as shown in **equation 4.6**.

$$\varepsilon_{\text{TE,sample}} = \frac{\Delta L_{\text{TE,sample}}}{L_{0,\text{sample}}} - \frac{\Delta L_{\text{TE,Zerodur}}}{L_{0,\text{Zerodur}}} \qquad [4.6]$$

---

^{c}Zerodur™ is a product of the company "Schott Glass Technologies".

4.1.2 Measurement uncertainty

In thermal expansion measurements there are several sources of measurement uncertainties. There are uncertainties in the temperature measurement, in the extensometer signal, in the calibration of the extensometers and in the determination of the measurement length. With values of 20 nm an 10.5 mK respectively, the resolutions of the extensometers and the temperature sensor are quite high. Because of this, the largest sources of uncertainties are the calibration and the determination of the measurement length. By repeated calibration the standard deviation on the calibration factor is determined to be about 0.5 %[d]. The typical measurement length is close to 50 mm. This length can be determined with an accuracy of about ±0.3 mm.

To determine the overall measurement uncertainties, the thermal expansion of a single sample is measured several times. Between each measurement the sample is removed, reinstalled and the extensometer is calibrated to account for all sources of uncertainties. The data of nine of such test runs is compared on the left of **figure 4.3**. The curves match very good with only small

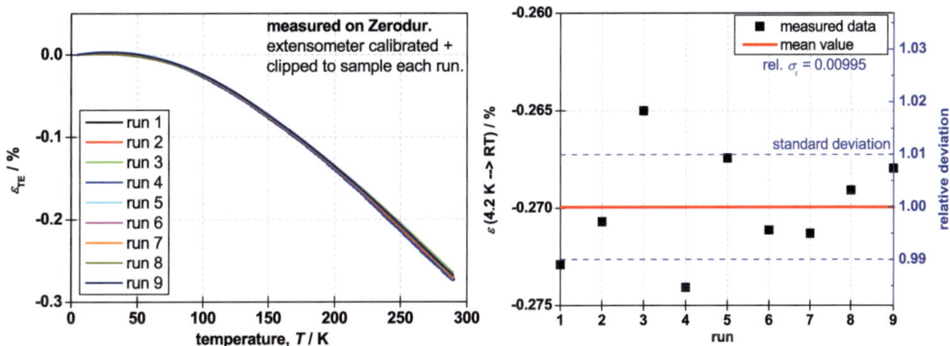

Figure 4.3: Overall measurement uncertainty of thermal expansion measurements. 9 test runs on Zerodur (left). Differences of the thermal expansion ε_{TE} (4.2 K → RT) of these test runs and the corresponding normalized standard deviation (right).

differences between all runs. The thermal expansion ε_{TE} from 4.2 K to room temperature is calculated for each test run. To get a quantitative understanding of the measurement uncertainties, ε_{TE} (4.2 K → RT) is compared for these test runs on the right of **figure 4.3**. A normalized standard deviation of $\sigma_{\varepsilon_{TE}} = 0.00995$ is obtained.

4.1.3 Characterized materials

In HTS cables, the thermal expansion is relevant in all material groups. Investigated materials are (grouped by material type):

---

[d]calculated from 9 calibrations. For each calibration, the sample is attached, detached and the measurement equipment is reset to account for all sources of errors.

Superconductor tapes

*RE*BCO tape SP-KIT-20100901, *RE*BCO bulk in ab-direction and Hastelloy[®e] C-276™.

Structural materials

stainless steel 316-LN, stainless steel Nitronic[®] 40 (S21900 / S21904)[f], aluminum Al-99.5 and copper.

Plastics

polyphenylsulfone (PPSU), polyether etherketone (PEEK), polyetherimide (PEI), polypropylene (PP), polycarbonate (PC), polyether sulfone (PES), polyamide (PA), polyphenylene ether (PPE) and polysulfone (PSU).

Composite materials

glass-fiber-reinforced-plastic G10 in normal and warp direction, polyurethanes (PUR) reinforced with stainless steel wires and Durotenax[®].

Insulation foils

polyimide foil.

Filling materials

Sn97Ag3 solder, In50Sn50 solder, In66Bi33.7 solder, Stycast[®g] blue (Stycast 2850 FT blue + Catalyst 9), Stycast[®] black (Stycast 2850 FT + Catalyst 9), epoxy resin (UHU plus endfest 300), epoxy resin Araldite (Araldite DBF + Aradur HY 951) and refined beeswax.

4.1.4 Results

The thermal expansion of *RE*BCO tapes from SuperPower (SP-KIT-20100901), *RE*BCO bulk in ab-direction (data from [ZLE05]) and Hastelloy C-276 (data from [LCZ08]) is shown on the left of **figure 4.4**. In SuperPower *RE*BCO tapes the *RE*BCO crystals are oriented with the ab-plane in tape direction. Comparing the thermal expansion of bulk *RE*BCO in ab-direction and SuperPower *RE*BCO tapes shows significant differences. In SuperPower *RE*BCO tapes the thermal expansion is mainly dominated by the substrate material used. Because of this, the thermal expansions of SuperPower tapes and Hastelloy C-276 match quite well with $\varepsilon_{\mathrm{TE}}^{RE\mathrm{BCO\ tapes}}(\mathrm{RT} \to 4.2\,\mathrm{K}) = -0.27\,\%$ and $\varepsilon_{\mathrm{TE}}^{\mathrm{C\text{-}276}}(\mathrm{RT} \to 4.2\,\mathrm{K}) = -0.29\,\%$. In an HTS cable, the superconducting tapes have to be protected from any damage during cool down and warm up. Thus, the thermal expansion

---

[e]"Hastelloy is the registered trademark name of Haynes International, Inc. The trademark is applied as the prefix name of a range of twenty two different highly corrosion-resistant metal alloys loosely grouped by the metallurgical industry under the material term 'superalloys' or 'high-performance alloys'." [Wika]

[f]"Nitronic 40 is a high-manganese stainless steel with high strength and excellent resistance to corrosion at high temperatures." [HP]

[g]"Stycast 2850 FT is a two component, thermally conductive epoxy encapsulant that can be used with a variety of catalysts. It features a low coefficient of thermal expansion and excellent electrical insulative properties." [Emea]

of the *REBCO* tapes are used as a reference and all other thermal expansions are compared with that reference.

4.1.4.1 Structural materials

On the right of **figure 4.4** the thermal expansions of the structural materials, stainless steel 316-LN, stainless steel Nitronic® 40, aluminum Al-99.5 and copper (data from [ZLE05]) are shown. Both stainless steel types, 316-LN with $\varepsilon_{\mathrm{TE}}^{316\text{-}LN}(\mathrm{RT} \to 4.2\,\mathrm{K}) = -0.29\,\%$ and Nitronic 40 with $\varepsilon_{\mathrm{TE}}^{\mathrm{Nitronic}\,40}(\mathrm{RT} \to 4.2\,\mathrm{K}) = -0.28\,\%$ match the thermal expansion of SuperPower *REBCO* tapes quite well and therefore are suitable candidates for the structure in an HTS cable. Pure aluminum Al-99.5 with $\varepsilon_{\mathrm{TE}}^{\mathrm{Al}\text{-}99.5}(\mathrm{RT} \to 4.2\,\mathrm{K}) = -0.41\,\%$ exhibits a much higher thermal expansion and should therefore not be used in combination with *REBCO* tapes. For example, the use of aluminum Al-99.5 results in significant mechanical stresses during and after cool down.

For thermal and electrical stabilization, copper is used in superconducting cables. With a thermal expansion of $\varepsilon_{\mathrm{TE}}^{\mathrm{Cu}}(\mathrm{RT} \to 4.2\,\mathrm{K}) = -0.32\,\%$ copper matches SuperPower *REBCO* quite reasonably.

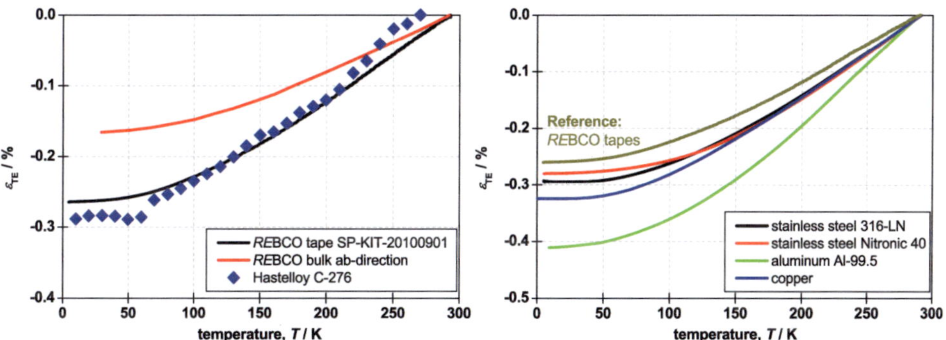

Figure 4.4: Thermal expansion of *REBCO* (left): *REBCO* tapes from SuperPower (SP-KIT-20100901), bulk *REBCO* in ab-direction and Hastelloy® C-276™. Thermal expansion of structural materials and electrical stabilization (right): stainless steel 316-LN, stainless steel Nitronic® 40, aluminum Al-99.5 and copper. Data from [ZLE05, LCZ08].

4.1.4.2 Insulating materials

High temperature superconductor cables also contain insulating materials. The thermal expansion of various plastic and composite materials are investigated. As shown on the left of **figure 4.5**, plastics exhibit high thermal expansions of $\varepsilon_{\mathrm{TE}}^{\mathrm{PEI}}(\mathrm{RT} \to 4.2\,\mathrm{K}) = -0.91\,\%$ to $\varepsilon_{\mathrm{TE}}^{\mathrm{PA}}(\mathrm{RT} \to 4.2\,\mathrm{K}) = -1.38\,\%$. The thermal expansion of composites is much lower. In the composites, the material which is added to the plastic dominates the overall thermal expansion. In glass-fiber-reinforced-plastic (G10) the thermal expansion is with $\varepsilon_{\mathrm{TE}}^{\mathrm{G10}\text{-}\mathrm{warp}}(\mathrm{RT} \to 4.2\,\mathrm{K}) = -0.24\,\%$ much lower in-plane with the glass mates or fibers (warp direction) than perpendicular to the glass mates or fibers (normal direction) with $\varepsilon_{\text{TE}}^{\text{G10-normal}}\,(\text{RT} \rightarrow 4.2\,\text{K}) = -0.60\,\%$. As shown on the right of **figure 4.5**, composite materials are preferable to pure plastics in HTS cables as composites match the thermal expansion of the superconductor tapes much more closely. In composites, by selecting the plastic, the reinforcement and the pattern the thermal expansion can be adjusted as needed. The reinforcing material has to be oriented in the direction of the superconductor tapes to avoid high differences in the expansions along the length of the HTS cable.

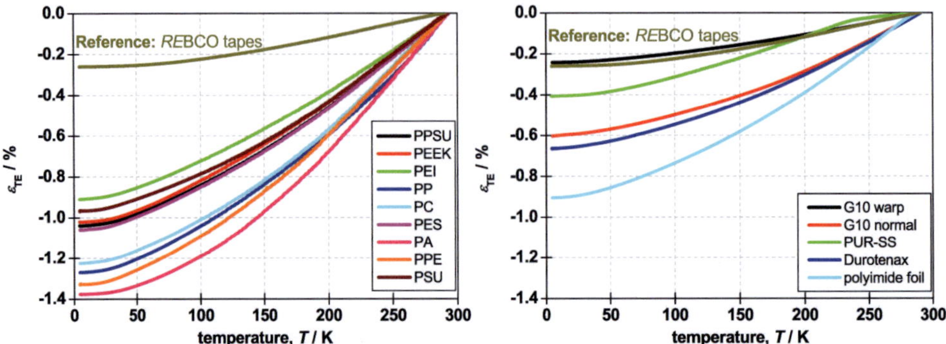

Figure 4.5: Thermal expansion of pure plastics (left): polyphenylsulfone (PPSU), polyether etherketone (PEEK), polyetherimide (PEI), polypropylene (PP), polycarbonate (PC), polyether sulfone (PES), polyamide (PA), polyphenylene ether (PPE) and polysulfone (PSU). Thermal expansion of composite materials (right): glass-fiber-reinforced-plastic (G10) in warp and in normal direction, polyurethanes reinforced with stainless steel wires (PUR-SS) and Durotenax. Data also published in [BGS 13].

To insulate superconducting cables polyimide foil is used. The thermal expansion of polyimide foil is $\varepsilon_{\text{TE}}^{\text{polyimide-foil}}\,(\text{RT} \rightarrow 4.2\,\text{K}) = -0.90\,\%$ and much higher than of *REBCO* tapes as shown on the right of **figure 4.5**. However, with thicknesses in the µm range and critical strains of more than 50 % [Dup], polyimide foil is very insensitive to mechanical strain and can be expected not to mechanically influence HTS cables. Because of this, polyimide foil can be used in HTS cables even though the mismatch of the thermal expansions of the foils and of *REBCO* tapes is exceptionally high.

4.1.4.3 Filling materials

In cable-in-conduit-conductors (CICC) of HTS cables, the tapes have to be connected between themselves and to the sheath of structural material to provide mechanical stabilization in high force environments [Kom95, p. 92][Wil83, p. 53]. This can be achieved with filling materials which fill any voids preventing movement of individual tapes. All tapes are connected, distributing mechanical loads evenly. Filling materials should be mechanically strong. However they have to match the thermal expansions of the *REBCO* tapes to avoid thermal stresses. To apply the filling materials with impregnation methods, their viscosities have to be close to liquids. Solders, glues or beeswax are possible candidates [Kom95, p. 92][Wil83, p. 313]. On the left of **figure 4.6**, the thermal expansions of Stycast blue, Stycast black, epoxy resin (UHU plus endfest 300 and Araldite) and refined beeswax are shown. Pure epoxy resins are not recommended due to their high thermal expansions of $\varepsilon_{\mathrm{TE}}^{\mathrm{UHU\ endfest}}\,(\mathrm{RT}\to 4.2\,\mathrm{K}) = -1.37\,\%$ respectively $\varepsilon_{\mathrm{TE}}^{\mathrm{Araldite}}\,(\mathrm{RT}\to 4.2\,\mathrm{K}) = -1.33\,\%$. The thermal expansions of Stycast black and Stycast blue are $\varepsilon_{\mathrm{TE}}^{\mathrm{Stycast\ black}}\,(\mathrm{RT}\to 4.2\,\mathrm{K}) = -0.50\,\%$ and $\varepsilon_{\mathrm{TE}}^{\mathrm{Stycast\ blue}}\,(\mathrm{RT}\to 4.2\,\mathrm{K}) = -0.53\,\%$ and are much closer to the expansion of the *REBCO* tapes. The thermal expansion mismatch is for beeswax with a thermal expansion of $\varepsilon_{\mathrm{TE}}^{\mathrm{beeswax}}\,(\mathrm{RT}\to 4.2\,\mathrm{K}) = -2.28\,\%$, exceptionally high.

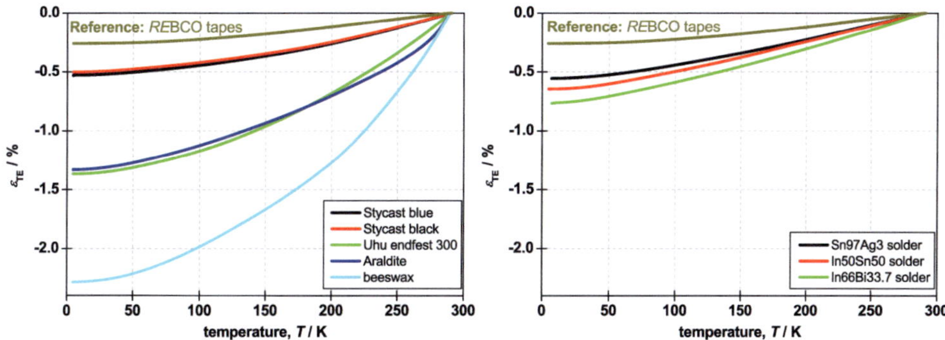

Figure 4.6: Thermal expansion of filling materials (glues, resins and beeswax on the left): Stycast blue, Stycast black, UHU plus endfest 300 epoxy resin, Araldite epoxy resin and refined beeswax. Thermal expansion of soft solders (right): Sn97Ag3 solder, In50Sn50 solder and In66Bi33.7 solder. Data published in [BBBW13].

The thermal expansions of soft solders are closer to *REBCO* tapes as shown on the right of **figure 4.6**. Their thermal expansion mismatches are with $\varepsilon_{\mathrm{TE}}^{\mathrm{Sn97Ag3}}\,(\mathrm{RT}\to 4.2\,\mathrm{K}) = -0.55\,\%$ for Sn97Ag3 solder, with $\varepsilon_{\mathrm{TE}}^{\mathrm{In50Sn50}}\,(\mathrm{RT}\to 4.2\,\mathrm{K}) = -0.65\,\%$ for In50Sn50 solder, and with $\varepsilon_{\mathrm{TE}}^{\mathrm{In66Bi33.7}}\,(\mathrm{RT}\to 4.2\,\mathrm{K}) = -0.77\,\%$ for In66Bi33.7 solder, still significant.

4.1.5 Summary and recommendation for HTS cables

The thermal expansions of materials for high temperature superconductor cables are summarized in **table 4.1**. SuperPower *REBCO* tapes are used as reference against which all other constituent materials of HTS cables (structure, insulation and filling materials) are compared.

Table 4.1: Thermal expansions from room temperature to 77 K and to 4.2 K of materials for HTS cables. Materials are grouped by their area of application: *RE*BCO tapes (**R**), structural materials (**S**), insulating materials (**I**) and filling materials (**F**).

	material	$\varepsilon_{\mathrm{TE}}\,(\mathrm{RT} \to 4.2\,\mathrm{K})$	$\varepsilon_{\mathrm{TE}}\,(\mathrm{RT} \to 77\,\mathrm{K})$	assessment
R	*RE*BCO tapes	$-0.27\,\%$	$-0.25\,\%$	reference for all materials
	C-276	$-0.29\,\%$	$-0.25\,\%$	very good match
S	316-LN	$-0.29\,\%$	$-0.28\,\%$	very good match
	Nitronic 40	$-0.28\,\%$	$-0.27\,\%$	very good match
	Al-99.5	$-0.41\,\%$	$-0.384\,\%$	better materials available
	copper	$-0.32\,\%$	$-0.30\,\%$	reasonable match
	pure plastics	$< -0.91\,\%$	$< -0.79\,\%$	unsuitable
I	G10 warp dir.	$-0.24\,\%$	$-0.22\,\%$	very good match
	G10 normal dir.	$-0.60\,\%$	$-0.53\,\%$	not recommended
	PUR-SS	$-0.40\,\%$	$-0.35\,\%$	reasonable match, can be improved by adjusting the ratios
	Durotenax	$-0.66\,\%$	$-0.59\,\%$	unsuitable
	Stycast blue	$-0.53\,\%$	$-0.47\,\%$	reasonable match, unsuitable in radiation
	Stycast black	$-0.50\,\%$	$-0.45\,\%$	reasonable match, unsuitable in radiation
F	epoxy resin	$-1.37\,\%$	$-1.24\,\%$	unsuitable
	Araldite	$-1.33\,\%$	$-1.20\,\%$	unsuitable
	In97Ag3 solder	$-0.55\,\%$	$-0.48\,\%$	significant deviation
	In50Sn50 solder	$-0.65\,\%$	$-0.55\,\%$	significant deviation
	In66Bi33.7 solder	$-0.77\,\%$	$-0.65\,\%$	significant deviation
	beeswax	$-2.28\,\%$	$-2.11\,\%$	unsuitable

Structural materials

For structural materials there are several choices. All tested stainless steel types matched the thermal expansion of the superconductor tapes very closely. Copper is also a reasonable match. Aluminum is not recommended as its thermal expansion is quite large.

Insulating materials

Pure plastics are unusable as insulation materials due to their large thermal expansion. Composites match the expansion of *RE*BCO tapes much more closely. Their thermal expansion can be matched to *RE*BCO tapes by choosing the correct base material, reinforcement and pattern. Composites are anisotropic and should be oriented with the direction which matches *RE*BCO tapes best in the direction of the cable.

Filling materials

As a filling material, Stycast blue or black matches the expansion of *RE*BCO tapes reasonably. However, in high neutron flux environments, e.g. in fusion magnets, these are unsuitable as Stycast degrades if exposed to radiation [Eva10, Emeb, Wil08]. Araldite epoxy resin is insensitive to neutron flux [WPSL12] and is approved for fusion applications [PHWM06]. The thermal expansion of Araldite is much higher than of Stycast. The large mismatch to *RE*BCO tapes renders pure Araldite unsuitable as filling material. Through the addition of sediments, e.g. sand, to glues and resins their thermal expansion can be influenced. Mixtures of sand and epoxy resin have been shown to exhibit much lower thermal expansions compared with pure epoxy resin [WWJ11]. By optimizing the mixing ratio it may be possible to match *RE*BCO tapes quite closely. The mixing of glues and resins with sediments is investigated in detail in **section 4.2**. There are also significant mismatches between the thermal expansions of *RE*BCO coated conductor tapes and of the soft solders Sn97Ag3, In50Sn50 and In66Bi33.7. With a low melting point of 117 - 126 °C and its low thermal expansion, In50Sn50 is the preferred solder.

4.2 Degradation free impregnation

To mechanically stabilize *RE*BCO tapes, voids have to be filled to prevent movement of the tapes and distribute loads equally. This can be done with glues, resins, solders or waxes. Due to their high mechanical strength, glues and resins can be assumed to provide excellent mechanical stabilization for the tapes. Impregnation with epoxy resin has already been tested in magnets of *RE*BCO tapes. Resin impregnated coils exhibited a strong decrease of their current carrying capabilities; this means that the *RE*BCO tapes themselves were degraded due to the resin impregnation [THT10, SBZ12]. *RE*BCO tapes are sensitive to perpendicular tensile stress (see **subsection 2.14**). In orientation perpendicular to the tape surface, stresses (25 MPa for *RE*BCO tapes from AMSC [LECS07]) still below the adhesion of glues and resins (up 50 - 80 MPa [WWJ11, MPV12, AP08]) can damage the tapes. During cool down, mismatches of the thermal expansions may therefore result in damage of the *RE*BCO tapes and degradation of their current carrying capabilities.

The thermal expansion of glues and resins is with $\varepsilon_{\mathrm{TE}}^{\mathrm{glues,\,resins}}\,(\mathrm{RT} \to 4.2\,\mathrm{K}) < -0.50\,\%$ in general much larger than the expansion of *RE*BCO tapes with $\varepsilon_{\mathrm{TE}}^{RE\mathrm{BCO\,tapes}}\,(\mathrm{RT} \to 4.2\,\mathrm{K}) = -0.27\,\%$ (see **table 4.1**). This means, to reduce the mismatch, the thermal expansion of the glues and resins has to be reduced drastically. Such a reduction can be achieved by adding fine grained sediments with very low thermal expansion such as glass or mineral powders.

Causes of and solutions for these degradations are investigated in three steps, which are published by the author in [BBW13]. In a first step the influence of the sediment content on the thermal expansion is examined for different glues and resins in **subsection 4.2.1**. Secondly in **subsection 4.2.2**, glues and resins with reduced thermal expansion are used to impregnate short *REBCO* samples. The current carrying capabilities, prior and after impregnation, are measured and the degradation calculated. Thirdly in **subsection 4.2.3**, a mixture of resin and sediment, which exhibits no degradation on short samples, is used to impregnate an HTS cable consisting of 15 *REBCO* tapes. The degradation of the cable is measured. The findings of this investigation are summarized in **subsection 4.2.4**.

4.2.1 Thermal expansion

In a first step the thermal expansions of mixtures of glues and resins with sediments are investigated.

Powder of amorphous silica dioxide (SiO_2), also referred to as fused silica or quartz has a very low thermal expansion of about $\varepsilon_{TE}^{fused\ silica}\,(RT \rightarrow 4.2\,K) \approx -0.015\,\%$ (extrapolated using data from [Quac]). Quarzgutmehl SILBOND® FW 12 EST a "electrically fused SiO_2 by iron-free grinding with subsequent air separation and coating with an organo-silicon compound" [Quad] is used as sediment in mixtures. It has an average grain diameter of $(17 \pm 9)\,\mu m$ and will be referred to as "quartz powder" in this work.

4.2.1.1 Characterized materials

Quartz powder is added to glues and resins using different mixing ratios. Samples for thermal expansion measurements were cast and the influence of the quartz powder content on the thermal expansion was determined. Mixtures with Stycast black (Stycast 2850FT + Catalyst 9), Stycast blue (Stycast 2850 FT + Catalyst 9) and Araldite epoxy resin (Araldite DBF + Aradur HY 951) are investigated. Stycast 2850 FT in combination with Catalyst 9, results in the lowest thermal expansion glue compared with the other catalysts, according to data published by the manufacturer [Emea]. Araldite epoxy resin is investigated because it is insensitive to neutron fluxes and is an approved material for fusion applications. The mixing ratios of the investigated mixtures are shown in **table 4.2**.

The viscosity of pure Araldite is higher than of pure Stycast black and pure Stycast blue. Because of this, Araldite allows much higher quartz powder contents, still resulting in mixtures with viscosity suitable for impregnation. Quartz powder contents of 14.29 % (weight) are the limit for Stycast, while Araldite remains usable up to quartz powder contents of 60 % (weight).

Table 4.2: Investigated mixtures of glues and resins with quartz powder (Quartzgutmehl SILBOND® FW 12 EST).

glue or resin	mixing ratios (resin : quartz powder)	quartz powder contents (weight)
Stycast black	pure, 24:1, 16:1, 12:1, 8:1 and 6:1	0 %, 4 %, 5.88 %, 7.69 %, 11.11 % and 14.29 %
Stycast blue	pure, 24:1, 16:1, 12:1, 8:1 and 6:1	0 %, 4 %, 5.88 %, 7.69 %, 11.11 % and 14.29 %
Araldite	pure, 4:1, 2:1, 1:1 and 2:3	0 %, 20 %, 33.33 %, 50 % and 60 %

4.2.1.2 Results

Stycast black (Stycast 2850 FT + Catalyst 9) exhibits the lowest thermal expansion of all tested filling materials as shown in **table 4.1**. The thermal expansion of pure Stycast black of $\varepsilon_{\mathrm{TE}}^{\text{pure Stycast black}} (\mathrm{RT} \to 4.2\,\mathrm{K}) = -0.50\,\%$ can be reduced through the addition of quartz powder down to $\varepsilon_{\mathrm{TE}}^{\text{Stycast black + quartz powder 6:1}} (\mathrm{RT} \to 4.2\,\mathrm{K}) = -0.41\,\%$. This matches the thermal expansion of *REBCO* tapes very closely. Regarding the thermal expansion, mixtures of Stycast black and quartz powder seem good choices as filling material for HTS cables. On the left of **figure 4.7**, the thermal expansion curves of mixtures with mixing ratios (Stycast black : quartz powder) of pure, 24:1, 16:1, 12:1, 8:1 and 6:1 are shown. The shape of these curves are similar; only the expansion is reduced by increasing the quartz powder content. On the right of **figure 4.7**, the influence of the quartz powder content on the thermal expansion room temperature to 4.2 K $\varepsilon_{\mathrm{TE}} (\mathrm{RT} \to 4.2\,\mathrm{K})$ and room temperature to 77 K $\varepsilon_{\mathrm{TE}} (\mathrm{RT} \to 77\,\mathrm{K})$ is shown.

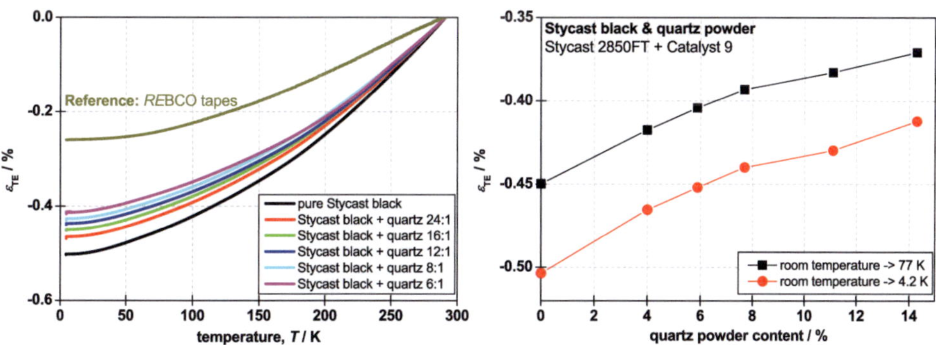

Figure 4.7: Thermal expansion of Stycast black (Stycast 2850FT + Catalyst 9) mixed with quartz powder (Quarzgutmehl SILBOND® FW 12 EST): thermal expansion curves (left). Dependence of thermal expansions $\varepsilon_{\mathrm{TE}} (\mathrm{RT} \to 4.2\,\mathrm{K})$ & $\varepsilon_{\mathrm{TE}} (\mathrm{RT} \to 77\,\mathrm{K})$ on quartz powder content (right).

Stycast blue (Stycast 2850 FT blue + Catalyst 9) is very similar to Stycast black. With $\varepsilon_{\mathrm{TE}}^{\text{pure Stycast blue}} (\mathrm{RT} \to 4.2\,\mathrm{K}) = -0.53\,\%$ the thermal expansion of Stycast blue is slightly higher than that of Stycast black. The same setup and mixing ratios are used as for Stycast black. As shown on the left of **figure 4.8**, the thermal expansion curves of Stycast blue are of similar shape. However, the curves are not as evenly spaced as in the Stycast black measurements. This results in a more uneven dependence of the thermal expansion of Stycast blue on the quartz powder content. These irregularities are caused by larger deviations between the assumed and actual mixing ratios of the Stycast blue mixtures. Measurement uncertainties of the thermal expansion measurements of the glue or resin mixtures are discussed in detail in **subsection 4.2.1.3**. The comparison of the thermal expansion from room temperature to $4.2\,\mathrm{K}$ $\varepsilon_{\mathrm{TE}}\,(\mathrm{RT} \to 4.2\,\mathrm{K})$ and room temperature to $77\,\mathrm{K}$ $\varepsilon_{\mathrm{TE}}\,(\mathrm{RT} \to 77\,\mathrm{K})$ between Stycast black and Stycast blue is shown on the right of **figure 4.8**. Regardless of quartz powder content, the thermal expansion of Stycast blue remains slightly larger than the expansion of Stycast black. Stycast black is therefore superior in this aspect.

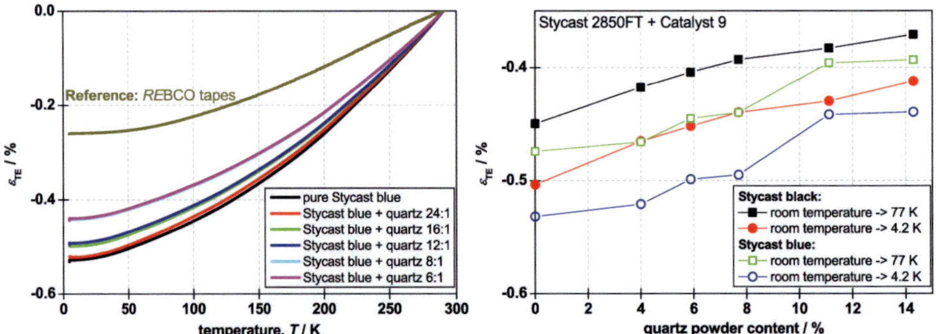

Figure 4.8: Thermal expansion curves of Stycast blue (Stycast 2850FT blue + Catalyst 9) mixed with quartz powder (Quarzgutmehl SILBOND® FW 12 EST) (left). Comparison of the thermal expansions $\varepsilon_{\mathrm{TE}}\,(\mathrm{RT} \to 4.2\,\mathrm{K})$ & $\varepsilon_{\mathrm{TE}}\,(\mathrm{RT} \to 77\,\mathrm{K})$ of Stycast black (Stycast 2850FT + Catalyst 9) and Stycast blue with increasing quartz powder content (right).

Araldite epoxy resin (Araldite DBF + Aradur HY 951) exhibits in pure form a large thermal expansion of $\varepsilon_{\mathrm{TE}}^{\mathrm{Araldite}}\,(\mathrm{RT} \to 4.2\,\mathrm{K}) = -1.33\,\%$. This is about 5 times higher than the expansion of SuperPower*REBCO* tapes. The large mismatch can be reduced by adding quartz powder to Araldite epoxy resin. At a mixing ratio (Araldite : quartz powder) of 2:3, which means a quartz powder content of $60\,\%$, the thermal expansion of the mixture is reduced to $\varepsilon_{\mathrm{TE}}^{\mathrm{Araldite\,+\,quartz\,powder\,2:3}}\,(\mathrm{RT} \to 4.2\,\mathrm{K}) = -0.51\,\%$. Higher quartz powder contents are not possible, they would result in viscosities unsuitable for casting and impregnation. At maximal mixing ratio, the addition of quartz powder resulted in a reduction of the thermal expansion of a factor of 2.6. The expansion of the mixture is quite close to the expansion of *REBCO* tapes. As shown on the left of **figure 4.9**, the thermal expansions curves of all Araldite and quartz powder mixtures are of similar shape. An increase of the quartz powder content reduces the thermal expansion almost linearly (**figure 4.9**, right).

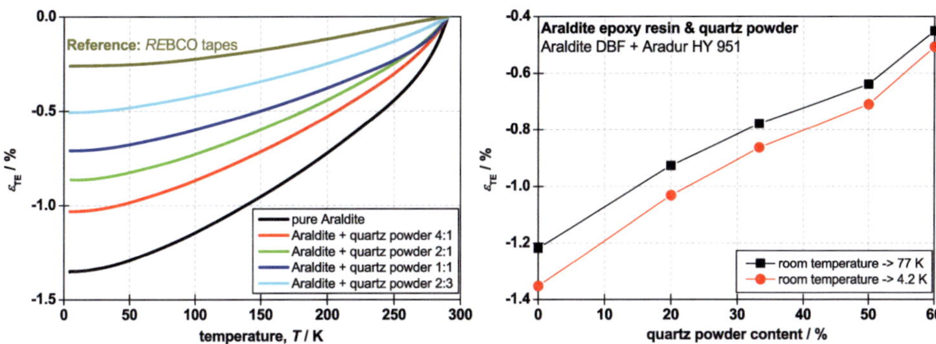

Figure 4.9: Thermal expansion of Araldite (Araldite DBF + Aradur HY 951) mixed with quartz powder (Quarzgutmehl SILBOND® FW 12 EST): thermal expansion curves (left). Dependence of thermal expansions $\varepsilon_{TE}\,(RT \to 4.2\,K)$ & $\varepsilon_{TE}\,(RT \to 77\,K)$ on quartz powder content (right).

The thermal expansions of pure Stycast and Araldite differ strongly, which can be drastically reduced with the addition of quartz powder at high mixing ratios. Due to the higher viscosity of pure Araldite, larger quartz powder contents are possible to reach critical viscosities.

In **table 4.3**, the thermal expansions from room temperature to 4.2 K $\varepsilon_{TE}\,(RT \to 4.2\,K)$ and room temperature to 77 K $\varepsilon_{TE}\,(RT \to 77\,K)$ are summarized for the mixtures with maximal quartz powder content.

Table 4.3: Thermal expansions of mixtures of glues and resins with quartz powder at maximal mixing ratios.

material	quartz powder	$\varepsilon_{TE}\,(RT \to 4.2\,K)$	$\varepsilon_{TE}\,(RT \to 77\,K)$	assessment
REBCO tapes	–	$-0.27\,\%$	$-0.25\,\%$	
Stycast black	$14.3\,\%$	$-0.41\,\%$	$-0.37\,\%$	good match, unsuitable in radiation
Stycast blue	$14.3\,\%$	$-0.44\,\%$	$-0.39\,\%$	good match, unsuitable in radiation
Araldite	$60\,\%$	$-0.51\,\%$	$-0.45\,\%$	reasonable match

At maximal quartz powder contents, mixtures with Stycast black, Stycast blue and Araldite seem suitable as filling materials for HTS cables with regards to their thermal expansions. As Stycast degrades if exposed to radiation [Eva10, Emeb, Wil08], Araldite and quartz powder mixtures are ideal filling materials in fusion application, e.g in cables of fusion magnets.

4.2.1.3 Measurement uncertainties

The overall uncertainties of the thermal expansion measurements of the glue and the resin mixtures consist of two parts:

First part: experimental error of the thermal expansion measurements

As elaborated upon in **subsection 4.1.2**, thermal expansion measurements are very precise. The normalized standard deviation of thermal expansion measurements is $\sigma_{\varepsilon,norm} = 0.00995$ of the measured value with the used setup. The thermal expansion measurement uncertainty corresponds to errors in the y-direction in **figure 4.7**, **figure 4.8** and **figure 4.9**. Nevertheless, since the deviations are less than 1 % of the measured values, they are neglected.

Second part: preparation of the samples

The preparation of samples is the major source of uncertainties. For each sample either 20 g of glue or of resin mixture are prepared. Depending on the desired mixing ratio (e.g. glue : quartz powder = 16:1), the corresponding amounts of glue or resin and quartz powder are calculated. Using scales with a 0.01 g resolution, these are added to a receptacle. Accounting for the accuracy of instruments, the environmental conditions (e.g. air currents, changes in pressure) and the skill of the experimenter, uncertainties of this step are assessed to be in the order of ± 0.03 g. Glue / resin and quartz powder are now thoroughly mixed and cast into a mold to form the sample for the thermal expansion measurement. The mixing leads to additional uncertainties. Poorly mixed quartz powder residues significantly influence the effective mixing ratios of the samples. Such deviations result in shifts in the x-direction in **figure 4.7** (graph on the right), **figure 4.8** (graph on the right) and **figure 4.9** (graph on the right). However, they cannot be specified quantitatively, for the mixing itself plays a major role. Therefore no x-direction error bars are shown in these figures. At low mixing ratios, especially in the Stycast samples, the x-direction deviations are presumably significant.

4.2.2 Degradation measurements on short *RE*BCO tapes

In a second step, the mismatches between the thermal expansions of *RE*BCO tapes and glues and resins have to be identified as cause of the degradation of epoxy resin impregnated *RE*BCO coils.

4.2.2.1 Measurement procedure

For these investigations, the current carrying capabilities of short *RE*BCO tapes are measured prior to and after impregnation with glues and resins of different thermal expansions. Copper stabilized *RE*BCO tapes from SuperPower (SP12050) which are punched to the meander structures necessary for Roebel-Assembled-Coated-Conductor (RACC) cables (see **section 5.2**) are used. Because of the punched out sections of the meander structure, at least one of the sides of the *RE*BCO tapes is always cut open. The copper stabilizer does not completely encase the tape. Compared with original tapes, meander-structured tapes can be expected to be more prone to mechanical damage and exhibit an increased sensitivity to perpendicular stresses. Additionally, at the open sides, the superconducting *RE*BCO layers are accessible from the outside. Due to this increased sensitivity, meander-structured tapes are a "worst case" scenario. If a meander-structured sample does not show any degradation it can be safely assumed that there will also be no degradation in the corresponding original tape.

All samples are prepared identically. Voltage taps are soldered to the tapes 10 cm apart. The critical current is measured in a liquid nitrogen bath (77 K) in self-field conditions. A critical electric field E_c of $1\,\mu V\,cm^{-1}$ is used to determine the critical currents. The tapes are cleaned with ethanol and are glued using the glue or resin mixture associated with that sample into a U-shaped former made of glass-fiber-reinforced-plastics (G10). The former has a total length of 9.5 cm. A G10 cap is attached and screwed to the former. Between the U-shaped former, the REBCO tape and the cap a space of 0.2 mm remains. The glue or resin mixture is distributed homogeneously by evacuating the sample. Excess glue or resin is removed. This setup is shown schematically in **figure 4.10**.

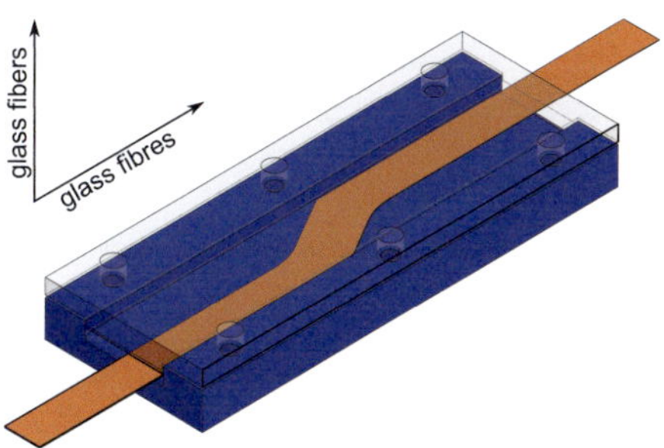

Figure 4.10: Schematic drawing of the setup used for degradation measurements on short REBCO tapes.

After 24 h curing time the current carrying capabilities are measured again using the same boundary conditions as before. The critical currents I_c are compared to determine the degradation occurring on this first cool down from room temperature to 77 K. At 4.2 K the mismatches between the thermal expansions are even larger than at 77 K. Therefore all samples are thermally cycled between room temperature and 4.2 K five times. In a liquid nitrogen bath their current carrying capabilities are measured again in self-field conditions.

With this measurement procedure three critical current values are determined (at self-field conditions at 77 K and a critical electric field of $1\,\mu V\,cm^{-1}$):

- The critical current I_c prior to glue or resin impregnation.

- The critical current I_c after glue or resin impregnation on first cool down to 77 K.

- The critical current I_c after glue or resin impregnation after repeated thermal cycling from room temperature to 4.2 K.

All values are compared and the degradations of the corresponding mixtures of glue or resin with quartz powder are calculated.

4.2.2.2 Results

Three different types of samples are made: samples using pure Araldite epoxy resin (Araldite DBF + Aradur HY 951), a mixture of Araldite epoxy resin and quartz powder with a mixing ratio (resin : quartz powder) of 1:1 and a mixture of Stycast black (Stycast 2850 FT + Catalyst 9) and quartz powder with a mixing ratio (Stycast : quartz powder) of 8:1. The quartz contents of these mixtures is slightly less than the maximal quartz powder contents of the corresponding glues or resins. The viscosities therefore remain high. This simplifies the impregnation of the tapes. All experiments are repeated twice, the data of both measurements is averaged.

In **table 4.4**, the normalized current carrying capabilities of the sample types are shown after impregnation and after repeated thermal cycling between room temperature and 4.2 K.

Table 4.4: Results of delamination measurements on short *REBCO* tapes. The normalized critical currents are shown after glue or resin impregnation and after repeated thermal cycling between room temperature and 4.2 K.

sample type	after impregnation		after thermal cycling	
	$I_c/I_{c,0}$	degradation	$I_c/I_{c,0}$	degradation
pure Araldite	0.77	22.6 %	0.04	95.8 %
Araldite + quartz powder 1:1	0.99	0.8 %	0.98	1.1 %
Stycast black + quartz powder 8:1	0.22	77.3 %	0.09	90.7 %

After impregnation, samples using pure Araldite resin exhibit significant degradation of their current carrying capabilities of 22.6 % on first cool down to 77 K. After repeated thermal cycling between room temperature and 4.2 K the degradation increases to 95.8 %. The tapes now hardly show any superconducting behavior. Samples impregnated with a mixture of Stycast black and quartz powder (8:1) are also strongly degraded. On first cool down to 77 K a degradation of 77.3 % occurs. This is increased by thermal cycling to 90.7 % degradation. Only the samples which are impregnated with a mixture of Araldite and quartz powder (1:1) carry the same current prior and after impregnation. On first cool down to 77 K, their current carrying capabilities are reduced by 0.8 %. Repeated thermal cycling between room temperature and 4.2 K hardly influences the critical current of this sample type. The degradation increases only slightly to 1.1 %.

This degradation can be prevented nearly completely by reducing the thermal expansion of the resin from $\varepsilon_{TE}^{\text{pure Araldite}}$ $(\text{RT} \rightarrow 4.2\,\text{K}) = -1.33\%$ to $\varepsilon_{TE}^{\text{Araldite + quartz powder 1:1}}$ $(\text{RT} \rightarrow 4.2\,\text{K}) = -0.70\%$ through the addition of quartz powder. Only the samples using this Araldite and quartz powder mixture do not exhibit degradation of the current carrying capabilities. Even though the thermal expansion of Stycast and quartz powder mixtures are lower, the corresponding samples degrade strongly. This can not be explained by thermal expansion mismatch.

As meander structured tapes, in which the superconducting layers are exposed on at least one side, are used in these experiments there may be other sources of degradation. Glues or resin may interact with the *RE*BCO in the superconducting layer causing other forms of degradation. To verify if this is the case with Stycast, a sample is made in which a very thin film of Stycast is applied only to the sides of the tape. In this sample, thermal expansion does not play any role. Due to the direct contact of Stycast and the superconducting *RE*BCO layer other forms of degradation remain possible. These samples exhibit a reduction of their current carrying capabilities to 0.12 of the nominal current. This is a degradation of 87.3 %. There seem to be an incompatibility between Stycast and *RE*BCO, degrading the current carrying capabilities on contact. Stycast is therefore unsuitable as for the impregnation of *RE*BCO tapes and coils and cannot be used as filling material in HTS cables.

On short *RE*BCO samples it has been shown that:

- Impregnation with pure Araldite leads to strong degradation of the current carrying capabilities of SuperPower *RE*BCO tapes. Cause of that degradation is the large mismatch of the thermal expansions.

- The degradation can be prevented by reducing the thermal expansion mismatch. Impregnation of SuperPower *RE*BCO tapes with a mixture of Araldite and quartz powder (mixing ratio 1:1) results in negligibly degradation of the critical current. This mixture is promising as filling material in HTS cables and coils.

- Pure Stycast, and any mixtures containing Stycast, are unsuitable to impregnate Super-Power *RE*BCO tapes. In these tapes, Stycast degrades the current carrying capabilities as it comes into contact with the superconducting *RE*BCO layer.

4.2.3 Validating results on an HTS cable

In a third step, the mixture of Araldite epoxy resin and quartz powder (1:1), which allowed degradation free impregnation of short samples is tested on an HTS cable. The sample is a 1.2 m long Roebel Assembled Coated Conductor (RACC) cable (see **section 5.2**) consisting of 15 *RE*BCO tapes (SuperPower SP12050 advanced pinning) which are punched to 5 mm width. It is one of the standard products of Industrial Research Limited (IRL) and is in the following referred to as a 15x5 RACC cable. Voltage taps are soldered to several tapes and to the copper terminations (or contacts) of the cable. The current carrying capabilities are determined in a liquid nitrogen bath (77 K) at self-field conditions. The voltage across the whole cable (at voltage taps at the copper terminations) is used to characterize the critical current of the cable. Ohmic contributions are subtracted. At $1\,\mu V\,cm^{-1}$ the critical current of the original cable is $I_{c,\,original}\left(1\,\mu V\,cm^{-1}\right) = 1935\,A$. At $5\,\mu V\,cm^{-1}$ the critical current is

$I_{c, \text{original}} \left(5\,\mu\text{V cm}^{-1}\right) = 2083\,\text{A}$. The n-values of the superconducting transition are calculated using the electric field range from the critical electric field ($E_c{=}1\,\mu\text{V cm}^{-1}$) to ten times the critical electric field ($10{\cdot}E_c{=}10\,\mu\text{V cm}^{-1}$) [TH05]. An n-value n_{orginal} of 21.5 is obtained for the whole cable.

Similar to the degradation measurements on short *REBCO* samples (**subsection 4.2.2**), the HTS cable is cleaned with ethanol and glued into a stabilizing jacket of glass-fiber-reinforced plastic (G10). As shown in **figure 4.11**, the jacket consists of two parts: a U-shaped former and a T-shaped cap. It nearly covers the complete distance between the copper terminations of the cable. A space of 1 mm remains between former RACC cable and cap for the impregnating resin.

Figure 4.11: Schematic drawing of a G10 jacket for the impregnation of a 15x5 RACC cable from IRL (shown in dark yellow). The jacket consists of two parts, a U-shaped former (shown in blue) and T-shaped cap (shown in transparent gray).

The impregnation is done by filling the former with the Araldite and quartz powder mixture, inserting the cable and adding additional resin mixture until the cable is completely covered. Using vacuum the resin is drawn into gaps and voids. Additional resin mixture is added to achieve complete coverage. This is repeated twice to get the Araldite and quartz mixture into the RACC cable. The jacket is closed with the T-shaped cap and the vacuum impregnation is repeated one last time. Excess resin is carefully removed. After 24 h curing at room temperature, the current carrying capabilities of the cable are measured again using the same boundary conditions. With the $1\,\mu\text{V cm}^{-1}$ criteria critical currents of $I_{c, \text{after impregnation}} \left(1\,\mu\text{V cm}^{-1}\right) = 1915\,\text{A}$ and with the $5\,\mu\text{V cm}^{-1}$ criteria of $I_{c, \text{after impregnation}} \left(5\,\mu\text{V cm}^{-1}\right) = 2090\,\text{A}$ are measured.

The critical current measurements prior to (original cable) and after the impregnation with the Araldite and quartz powder mixture are compared on the left of **figure 4.12**.

The electric field vs. current curves are nearly identical. After impregnation, the normalized current carrying capabilities are:

- 0.990 with the $1\,\mu\text{V cm}^{-1}$ criteria. This corresponds to a degradation of the current carrying capabilities of 1.03 %.

- 1.003 with the $5\,\mu\text{V cm}^{-1}$ criteria. This corresponds to increase of the current carrying capabilities of 0.33 %.

Within the measurement uncertainty, that impregnation of the 15x5 RACC cable with a mixture of Araldite epoxy resin and quartz powder (1:1) does not result in any detectable degradation of the current carrying capabilities.

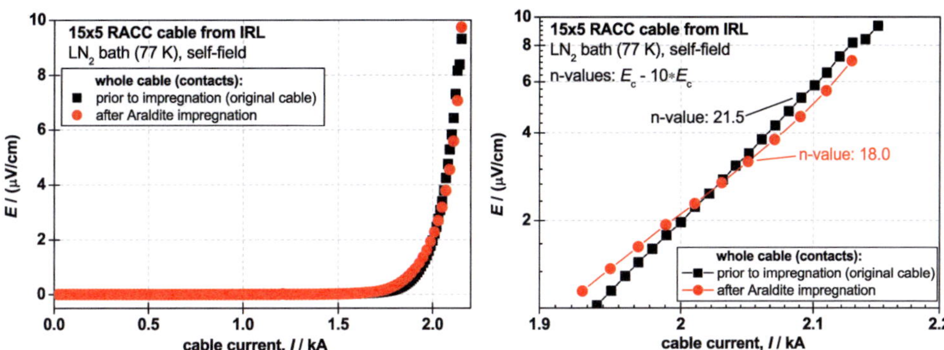

Figure 4.12: Electric field vs. current of a 15x5 RACC cable from IRL at 77 K in self-field conditions for the voltage taps at the copper terminators. Ohmic contribution are subtracted. Data is shown for the original cable (in black) and after impregnation with a mixture of Araldite and quartz powder (in red). Linear x-axis and y-axis (left) and logarithmic x-axis and y-axis (right).

The n-value of the superconducting transition of the impregnated cable is calculated in the same electric field range as for the original cable (range E_c to $10 \cdot E_c$). An n-value of $n_{\text{after impregnation}} = 18.0$ is obtained as shown on the right of **figure 4.12** with logarithmic x-axis and y-axis scaling. This indicates a marginal reduction in steepness of the curve of the superconducting transition. However, due to the G10 former and the epoxy impregnation, the boundary conditions of the original and the impregnated cable differ. In more detail, the original cable is in direct contact with the liquid nitrogen bath while the impregnated cable is thermally insulated with G10 and epoxy resin. During the superconducting transitions, the temperature of the impregnated cable may increase, whereas the original cable remains at 77 K due to the more effective cooling contact surface. Thus, the test boundary conditions of these two samples are not identical, limiting the significance of a comparison of n-values.

The 1.2 m HTS cable behaves similarly to the short *REBCO* samples impregnated with the same mixture (see **table 4.4**). The mixture of Araldite epoxy resin and quartz powder (1:1) can be recommended as filling material for HTS cables and in wet wound coils.

4.2.4 Summary

In high force environments filling materials are necessary to mechanically stabilize HTS cables. With fillers, movement of individual tapes is prevented. The tapes are connected and mechanical loads are distributed evenly. Due to their high mechanical strength and easy applicability, glues and resins are preferred filling materials. In *REBCO* coils, epoxy resin impregnation has been shown to mechanically stabilize the tapes while causing strong degradation of their current carrying capabilities [THT10, SBZ12]. The large mismatches of the thermal expansions of the resins and the *REBCO* tapes are expected to be the cause of these degradations. This is investigated in a three step approach:

- Firstly it has been shown that the thermal expansion of glues and resins can be adjusted through the addition of sediments. With a fine grained powder from amorphous silica dioxide (quartz powder: SILBOND® FW 12 EST) the thermal expansion of Araldite epoxy resin (Araldite DBF + Aradur HY 951) is reduced from $\varepsilon_{TE}^{\text{pure Araldite}}(\text{RT} \rightarrow 4.2\,\text{K}) = -1.33\%$ to $\varepsilon_{TE}^{\text{Araldite + quartz powder 2:3}}(\text{RT} \rightarrow 4.2\,\text{K}) = -0.51\%$ while retaining a viscosity suitable for casting and impregnation. With Stycast black (2850 FT + Catalyst 9) a reduction from $\varepsilon_{TE}^{\text{pure Stycast black}}(\text{RT} \rightarrow 4.2\,\text{K}) = -0.50\%$ to -0.41% is possible at maximal mixing ratio (14.3 % sediment). The thermal expansions of the mixtures are closer to $\varepsilon_{TE}^{REBCO\text{ tapes}}(\text{RT} \rightarrow 4.2\,\text{K}) = -0.27\%$ which is the thermal expansion of SuperPower REBCO tapes.

- Secondly the current carrying capabilities of REBCO tapes are measured prior to and after impregnation with glues and resins of different thermal expansions. To increase the mechanical sensitivity of the tapes, all samples are punched to meander structures as needed for HTS Roebel cable assembly (see **section 5.2**). With pure Araldite epoxy resin, SuperPower REBCO tapes (SP12050) degrade by 22.6 % on first cool down. The degradation increases to 95.8 % after repeated thermal cycling between room temperature and 4.2 K. With a mixture of Araldite and quartz powder of the mixing ratio (resin : quartz powder) of 1:1 only small degradations occur: 0.8 % on first cool down and 1.1 % after thermal cycling. The current carrying capabilities of Stycast impregnated tapes degrade almost completely. 77.3 % on first cool down and 90.7 % after thermal cycling. As the thermal expansion of Stycast is lowest of all tested glues and resins, this can not be contributed to thermal expansion mismatch. Degradation is also observed in samples with a thin film of Stycast was applied solely to the punched open sides. Here the superconducting REBCO layer of the tape is accessible from the outside. There seems to be an incompatibility between Stycast and REBCO which is cause of this degradation. In all tested samples, only the mixture of Araldite epoxy resin and quartz powder (mixing ratio 1:1) does not degrade the current carrying capabilities of the REBCO tapes.

- Thirdly, the mixture of Araldite epoxy resin quartz powder (mixing ratio 1:1), which showed no degradation on the short samples, is tested on a 1.2 m long 15x5 RACC cable. The current carrying capabilities of the cable are determined prior to and after impregnation. Over the whole cable, using the voltage taps at the copper terminations, no degradation is observed.

Therefore, thermal expansion mismatch is at least one of the causes of the degradation of epoxy impregnated REBCO coils. A mixture of Araldite epoxy resin and quartz powder (mixing ratio 1:1) exhibits a reduced thermal expansion, much closer to REBCO tapes. With this mixture, degradation free impregnation of *REBCO* tapes and cables is possible. The mixture is recommended as filling material for HTS cables.

In addition, the thermal conductivities of the materials have to be investigated to avoid failures of the cooling of the superconducting cables.

4.3 Thermal conductivity

Temperature gradients result in heat transfer through microscopic diffusion and collisions of particles and quasi particles. In solid state matter heat is transferred through the body itself. Two carriers are dominant: free electrons and vibrations of the lattice called phonons. The overall thermal conductivity λ of a solid state body is the sum of the phononic λ_p and electronic thermal conductivity λ_e as shown in **equation 4.7**.

$$\lambda = \lambda_e + \lambda_p \qquad [4.7]$$

- In metals the electrons are not localized. They are not bound to specific atoms of the lattice and can move freely without interacting between themselves [Dru00]. This quantum mechanical version of an ideal gas is referred to as the free electron gas of metals [Som28]. The mobility of the electrons is very high. As electrons carry charge, metals in general are good electrical conductors. This also means, as heat is transferred through electrons, the electronic part of the thermal conductivity λ_e is dominant in metals [WF53] and depends on the temperature. At low temperature the movability of the electrons is high. Increasing the temperature results in higher numbers of phonons. As electrons are scattered at phonons, increasing the temperature reduces the mean free path of the electrons. This limits their movability and reduces the thermal conductivity.

- In insulators, electrons are in the valence band and are separated from the conduction bands by energy gaps. Because of this, the average mobility of the electrons is very low at room temperature and below. At these temperatures, the electrical conductivity and electronic thermal conductivity λ_e are low as well. In insulators, the thermal conductivity is dominated by the phononic contribution λ_p which depends on temperatures, too. At higher temperatures, there are more phonons and the bound electrons are excited thermally which increases their chances to break free of their covalent bond. In general, higher temperatures result in higher thermal conductivities in insulators.

The thermal transport depends on the temperature difference, the cross-section area, the distance and the thermal conductivity of the material. Therefore in the SI unit system thermal conductivity is given in Watt per Kelvin meter ($\frac{W}{K\,m}$).

In high temperature superconductor cables stable operating temperatures are required. Heat which is generated in the superconductor, through ohmic contributions or AC losses, has to be conducted to the thermal sink without exceeding the operating temperature limits. Thus, the thermal conductivity of the used materials is important in designing the cooling system. As the thermal conductivity depends on the material, the direction and the temperature, an in depth investigation is essential.

4.3.1 Experimental setup

Thermal conductivity is measured using the P670 Thermal Transport Option (TTO) [Quab] of the Physical Property Measurement System (PPMS$^{®}$) from QuantumDesign [Quaa]. This is a commercial system which determines the thermal conductivity at low temperatures by axial heat fluxes at a stationary temperature equilibrium [Sch09, p. 31]. One side of a sample is connected to a thermal sink while the other side is heated with a defined heating power. This results in a temperature gradient along the sample. If this gradient does not change with time, the sample is in static thermal equilibrium. The heat flux per cross section area ($\frac{Q}{A}$) is proportional to the temperature gradient ($\frac{dT}{dx}$).

$$\frac{Q}{A} = -\lambda^\star \cdot \frac{dT}{dx} \qquad [4.8]$$

In an ideal system, without any other heat fluxes e.g. through radiation or convection, the negative value of the constant of proportionality of **equation 4.8** corresponds to the thermal conductivity λ of the used material.

$$\text{ideal system:} \quad \lambda = \lambda^\star$$

In the Thermal Transport Option (TTO) of the Physical Property Measurement System (PPMS) the temperature gradient ΔT is determined with two calibrated Cernox temperature sensors which are attached to the sample, a short distance away from the cold and warm ends. The distance between the temperature sensors is the measurement distance Δx. This is a thermal four point measurement which improves the accuracy comparable to electrical four point measurements. The whole measurement setup is inclosed in a radiation shield. This shield is kept a the same temperature as the temperature sink on the cold end of the sample to minimize radiation losses. In addition, the sample chamber is pumped to a vacuum of the order of 10^{-6} mbar [Sch09, p. 33] and thin manganin[h] wires are used for the instrumentation and the heater to reduce heat flux through convection and conduction. In this system the thermal conductivity of the sample is given with **equation 4.9**.

---

[h]"Manganin is a trademarked name for an alloy of typically 86 % copper, 12 % manganese, and 2 % nickel." [Wikb]

$$\text{PPMS:} \quad \lambda = \left(\frac{Q - Q_{\text{rad}}}{T_{\text{hot}} - T_{\text{cold}}} - C_{\text{wires}} \right) \cdot \frac{\Delta x}{A} \tag{4.9}$$

In **equation 4.9**, Q corresponds to the heating power, Q_{rad} is the radiation loss, and C_{wires} is the heat loss at the instrumentation and the heater wires while A is the cross sectional area of the sample.

- Radiation losses Q_{rad} can be approximated using the Stefan–Boltzmann law. Heat is radiated from the sample. The outgoing heat flux $j^{\star}_{\text{out}}$ is directly proportional to the product sample area A, emissivity ε and fourth power of sample temperature T^4_{sample} (**equation 4.10**). At the same time, there is in ingoing heat flux $j^{\star}_{\text{in}}$ from the environment to the sample. $j^{\star}_{\text{in}}$ is proportional to the sample area A, emissivity ε and fourth power of the temperature of the environment $T^4_{\text{enviroment}}$ (**equation 4.11**). In thermal equilibrium, the radiation losses are the difference of outgoing and incoming heat flux (**equation 4.12**). Within the PPMS, approximately half the sample is at the high temperature assuming a uniform temperature gradient. Therefore, the radiation losses can be approximated as the product of Stefan-Boltzmann constant $\sigma = 5.670\,400 \times 10^{-8}$ J/(s m^2 K^4), the emissivity of the sample ε, half of the surface $\frac{O}{2}$, and the difference of the fourth powers of the temperatures at the hot end of the sample and the thermal sink $\left(T^4_{\text{hot}} - T^4_{\text{sink}} \right)$ (see **equation 4.13**).

$$j^{\star}_{\text{out}} = \sigma \cdot A \cdot \varepsilon \cdot T^4_{\text{sample}} \tag{4.10}$$

$$j^{\star}_{\text{in}} = \sigma \cdot A \cdot \varepsilon \cdot T^4_{\text{enviroment}} \tag{4.11}$$

$$Q_{\text{rad}} = j^{\star}_{\text{out}} - j^{\star}_{\text{in}} \tag{4.12}$$

$$\approx \sigma \cdot \frac{O}{2} \cdot \varepsilon \cdot \left(T^4_{\text{hot}} - T^4_{\text{sink}} \right) \tag{4.13}$$

- Heat losses due to heat conduction in the instrumentation and heater wires have been analyzed empirically. These losses are described by the correction polynom C_{wires}. The coefficients of **equation 4.14** have been determined empirically and are given by the manufacturer of the measurement system. $\alpha = 1.995 \times 10^{-8}$ W K^{-2}, $\beta = 9.555 \times 10^{-10}$ W K^{-3} and $\gamma = 1.008 \times 10^{-11}$ W K^{-4} [Sch09, p. 32].

$$C_{\text{wires}} \approx \alpha \cdot T + \beta \cdot T^2 + \gamma \cdot T^3 \tag{4.14}$$

- Heat losses due to convection can be neglected due to the high vacuum of the order of 10^{-6} mbar in the sample chamber.

This method to measure thermal conductivity is shown schematically in **figure 4.13**.

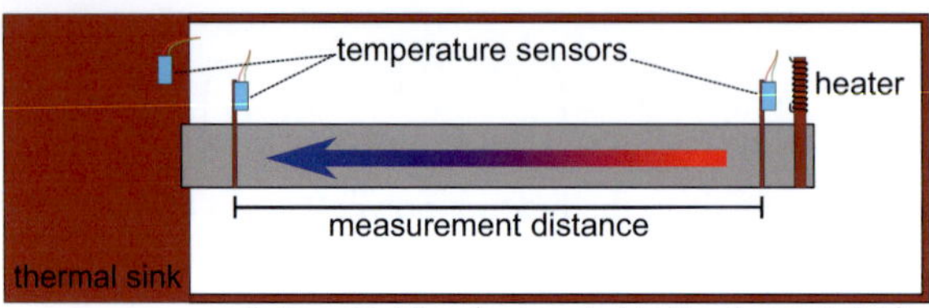

Figure 4.13: Schematic drawing of the measurement setup for thermal conduction measurements. A sample is heated with a defined heating power on one side while the other side is connected to a thermal sink and thus kept at constant temperature. The temperature gradient is measured with two high precision temperature sensors. The sample chamber is surrounded by a radiation shield which is connected to the thermal sink.

4.3.2 Measurement uncertainty

Thermal conductivity λ is determined using **equation 4.9** and the corrections from **equation 4.13** and **equation 4.14**. The cross section area A, the measurement distance Δx, the temperatures T_{hot}, T_{cold}, T_{sink}, the heating power Q and the emissivity ε are major sources of uncertainties.

Geometry

In samples with an easy geometric shape, e.g cylinders or rectangular prisms, the combined uncertainties of the determination of the cross section area A and the measurement distance Δx are about 3 - 5 %. In uncommon geometries this can increase up to 10 %. Geometries errors can be avoided by measuring samples of the same material using slightly different shapes [BGS13].

Temperature

The temperatures, T_{hot}, T_{cold}, T_{sink} are measured with calibrated high precision Cernox sensors. Uncertainties of the temperature measurements are about 2 % at cryogenic temperatures and 1 % at room temperature[i]. The supply current of the temperature sensors is adjusted automatically depending on the temperature. Low supply currents are used at cryogenic temperatures, therefore self-heating of the sensors is negligible [BGS13].

Heating power

The power of the heater can be controlled very precisely. According to the manufacturer a minimal current of $10\,\mu\text{A}$ can be kept within $\pm0.02\,\mu\text{A}$. This results in a maximal uncertainty of the heating power of about 0.23 % [Sch09, p. 40].

Radiation losses

The emissivity ε of samples is in range of $0.1 > \varepsilon > 1$ depending on the material and the surface conditions. The emissivity of a sample can only be estimated very roughly. As the radiation losses are proportional to the product of emissivity and the difference of the fourth powers

---

[i]Estimated using manufacturer data of the temperature sensors [Laka].

of the temperatures at the hot end of the sample and the thermal sink, uncertainties of the emissivity become dominant at higher temperatures. For samples with high thermal conductivity the uncertainty of the radiation losses is usually below 1 % at cryogenic temperatures. From 200 to 300 K this increases from 5 to 7 % [Bag12, BGS13]. For samples with a low thermal conductivity, e.g. insulators, this contribution can be much larger and the uncertainty can increase up to 100 % at room temperature [Sch09, p. 41].

To summarize, thermal conductivity measurement uncertainties depend on the material, the geometry of the sample and the temperature range. At cryogenic temperatures error bars are often not even visible. Near room temperature, the measurement uncertainty of this method is within range of several ten percent.

To optimize the resolution of thermal conductivity measurements suitable sample geometries are essential: long and thin samples for good thermal conductors and wide and short samples for thermal insulators.

4.3.3 Characterized materials

For stable operation of superconducting magnets, heat has to be extracted from the cable without exceeding the temperature margin. To design a cooling system, the thermal conductivity of all materials in the cable is of interest. Materials, which are suitable choices (e.g. with low thermal expansion mismatches) in their corresponding areas of application, are investigated. The investigated materials are (grouped by material type):

Superconductor tapes

REBCO tapes SCS12050 parallel and perpendicular to the tape, *REBCO* tapes SF12050 parallel and perpendicular to the tape.

Structural materials

stainless steel 316-LN.

Composite materials

glass-fiber-reinforced-plastic G10 in normal and warp direction.

Insulation foils

polyimide foil.

Filling materials

In50Sn50 solder as it has a low thermal expansion and a low melting point of 116 - 127 °C, Stycast® blue (Stycast 2850 FT blue + Catalyst 9), Araldite epoxy resin and a mixture of Araldite 45.5 % quartz powder.

4.3.4 Results

*RE*BCO tapes are highly anisotropic materials. As shown in **subsection 2.3.3.2**, their mechanical properties and their in-field performance depend strongly on the orientation. Due to the layered layout of *RE*BCO tapes (see **figure 2.13**) the thermal conductivity is also anisotropic and varies substantially between parallel and perpendicular orientation.

4.3.4.1 *RE*BCO tapes, parallel to the tape

The thermal conductivity of *RE*BCO tapes in parallel orientation can be measured in a four point orientation. A heater is attached to one side of a *RE*BCO tape. This side of the tape is heated with a defined heating power. The other side of the tape is connected to a thermal sink. In thermal equilibrium a temperature gradient stabilizes. The temperature gradient is determined with two Cernox temperature sensors which are attached to temperature taps soldered onto the tape. The distance between the temperature taps is the measurement distance Δx.

In parallel orientation, the layers of the *RE*BCO tapes can be considered as thermal resistors connected in parallel. In this orientation the layer with the lowest thermal resistance, thus the highest thermal conductivity, dominates. In copper stabilized *RE*BCO tapes this is the stabilizing copper layer. The measurement setup and the corresponding equivalent circuit diagram of the thermal resistances is shown in **figure 4.14** [Sch09].

Due to the high thermal conductivity of the stabilizing copper layer, the thermal conductivity of copper stabilized tapes (SCS12050) is several decades higher than of stabilization-free tapes (SF12050). As shown on the left of **figure 4.16** the stabilized tapes show a distinct peak of about $835\,\mathrm{W\,K^{-1}\,m^{-1}}$ at 22 K. The copper used for the stabilizing layer must therefore be very pure, with a residual resistivity ratio (RRR) of $50 - 100$. The measurement uncertainty is small (few percent range) due to the four point configuration, high thermal conductivity of the sample and large measurement distance Δx of several cm.

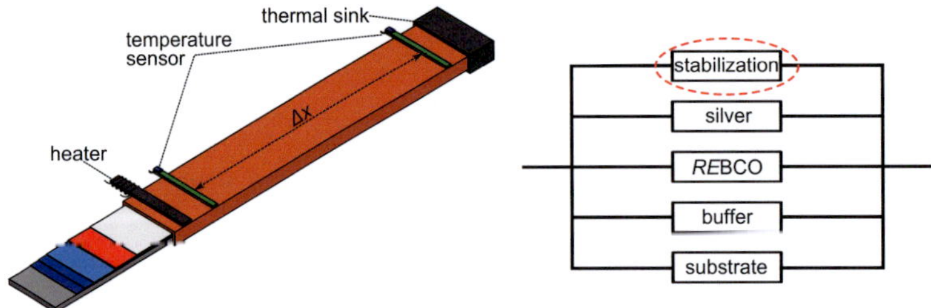

Figure 4.14: Measuring the thermal conductivity of *RE*BCO tapes parallel to the tape. A temperature gradient is generated with a heater and a thermal sink. The gradient over the measurement distance Δx is measured with two temperature sensors. In this orientation the layers can be represented as thermal resistors in parallel connection. The layer with the lowest thermal resistance, the stabilizing layer dominates. Picture after [Sch09].

4.3.4.2 *RE*BCO tapes, perpendicular to the tape

In perpendicular orientation, due to the very thin aspect ratio of *RE*BCO tapes the measurement distance over a single tape is far too short. Several tapes are soldered into a stack to increase the measurement distance Δx to a usable level. In50Sn50 solder is used to connect the tapes and form the stack. The contribution of the solder is subtracted from the measurement data. Copper plates are soldered to both sides of the stack. The heater and one of the two Cernox temperature sensors are attached to one of the copper plates. The other plate is connected to the thermal sink and to the second temperature sensor. This is a thermal two point measurement setup.

In this orientation the layers of the YBCO tapes can be represented as thermal resistors connected in series. Thus layers with high thermal resistance, which means low thermal conductivity, dominate. With copper stabilized tapes from SuperPower the (SCS12050 or SCS4050) the copper completely surrounds the tapes. The copper on the sides is parallel with the layered structure of the tape. In the equivalent circuit diagram the thermal resistor representing the copper stabilization on the sides of the tapes is therefore connected in parallel. To determine the influence of this parallel thermal connection, the copper on the sides of stabilized *RE*BCO tapes (SCS12050) can be punched away. This eliminates the parallel thermal resistor "stabilization on sides" as shown schematically in **figure 4.15**.

The buffer and *RE*BCO layers are comparable to insulators. Their thermal conductivity is very low. Therefore the overall thermal conductivity of *RE*BCO tapes is very low perpendicular to the tape. As shown on the right of **figure 4.16**, maximal thermal conductivities of $3.3\,\mathrm{W\,K^{-1}\,m^{-1}}$ for stabilized tapes (SCS12050) and $2.3\,\mathrm{W\,K^{-1}\,m^{-1}}$ for unstabilized tapes (SF12050) are gained. With less than $1\,\mathrm{W\,K^{-1}\,m^{-1}}$ the difference between SCS12050 and SF12050 is very low compared with the huge impact of the copper stabilization in parallel orientation. The impact of the copper on the sides of stabilized tapes is negligible due to the very low cross section area. The thermal conductivities of SCS12050 tapes and stabilized tapes with punched away copper sides (SCS12050 - punched) are nearly identical. Both curves are within one standard deviation of each other for almost the whole temperature range.

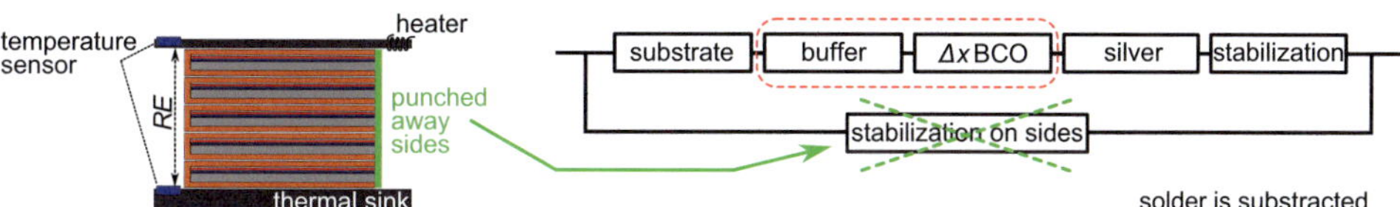

Figure 4.15: Measuring the thermal conductivity of *RE*BCO tapes perpendicular to the tape. Several tapes are soldered into a stack to gain a usable measurement distance Δx. A heater and temperature sensor are connected on one side of the stack while the other side is connected to the thermal sink and another temperature sensor. In this orientation the thermal resistances of the layers are connected in series. High thermal resistance layers are dominating. The Cu stabilization completely surrounding the tape is a parallel thermal connection which can be removed by punching away the sides of such tapes. The influence of the solder is subtracted.

Due to the two point measurement configuration, the low measurement distance Δx and the low thermal conductivity of the samples, measurement uncertainties are significant. Visible as error bars on the right of **figure 4.16**, the measurement uncertainties increase up to $\pm 0.27 \, \mathrm{W\,K^{-1}\,m^{-1}}$ at room temperature.

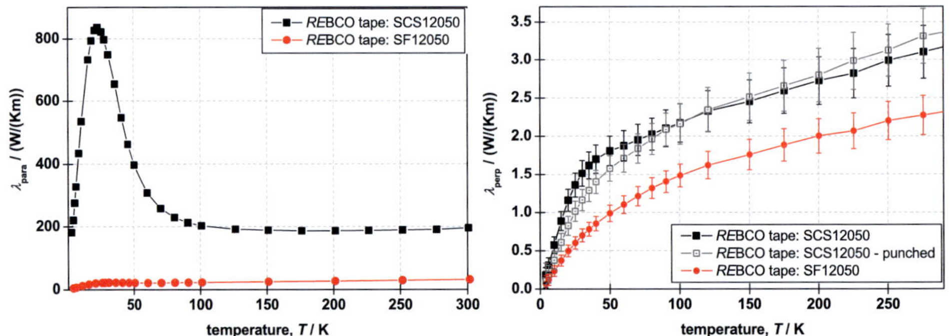

Figure 4.16: Thermal conductivity of *RE*BCO tapes in parallel (left) and perpendicular (right) orientation. Data for 12 mm wide copper stabilized tapes (SCS12050) and stabilization free tapes (SF12050) from SuperPower is shown. Data from [BGJW12].

4.3.4.3 Structural and insulating materials

To withstand Lorentz forces in a magnet, *RE*BCO tapes of high temperature superconductor cables have to be mechanically stabilized with structural materials. Due to excellent mechanical properties at cryogenic temperatures, austenitic stainless steels are the preferred choice. With a yield strength of 1015 MPa, tensile strength of 1701 MPa at 4.2 K [MKT93] and a low thermal expansion of $\varepsilon_{\mathrm{TE}}^{\mathrm{316\text{-}LN}}(\mathrm{RT} \to 4.2\,\mathrm{K}) = -0.29\,\%$ (see **figure 4.4**) 316-LN is an ideal structural material for HTS cables. As shown on the left of **figure 4.17**, the thermal conduction of 316-LN $\lambda^{\mathrm{316\text{-}LN}}$ increases with temperature from $\lambda^{\mathrm{316\text{-}LN}}(4.2\,\mathrm{K}) = 0.25\,\mathrm{W\,K^{-1}\,m^{-1}}$ at 4.2 K to $\lambda^{\mathrm{316\text{-}LN}}(77\,\mathrm{K}) = 6.8\,\mathrm{W\,K^{-1}\,m^{-1}}$ at 77 K. For 316-LN, there is no local maximum of the thermal conduction in the 20 K range as it is typical for pure electrical conductors e.g. copper or silver. However, the overall thermal conductivity of 316-LN is still 3 to 20 times higher than of insulators or *RE*BCO tapes in perpendicular orientation.

The thermal conductivity of composite materials as glass-fiber-reinforced-plastics G10 is anisotropic. For the thermal conductivity the anisotropy of G10 is much less pronounced compared with the thermal expansion. Warp (parallel to the glass fibers) and normal direction (perpendicular to the glass fibers) behave quite similarly. The thermal conductivity of G10 in warp and in normal direction increases monotonously with temperature. Regardless of temperature, λ^{G10} is always higher in warp direction than in normal direction by a fac-

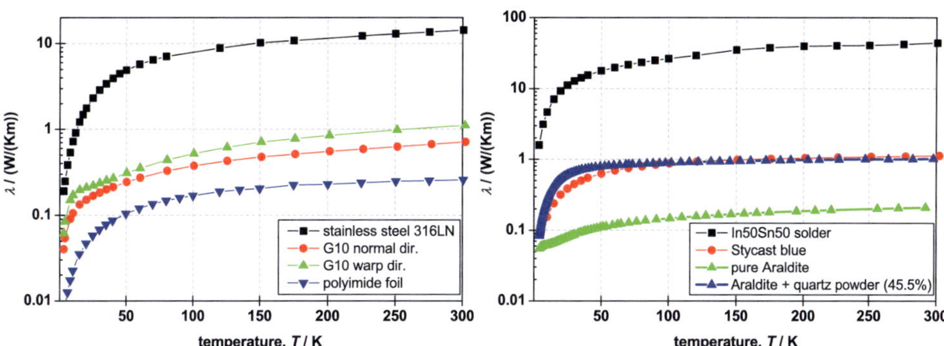

Figure 4.17: Thermal conductivity of structural materials, insulating materials suitable for HTS cables (left). Thermal conductivity of filling materials and In50Sn50 soft solder (right). Data from [BGS13, GR74].

tor of 1.2 to 2. In the temperature range of 4.2 K to 77 K the thermal conductivity changes from $\lambda^{\text{G10-warp}}(4.2\,\text{K}) = 0.08\,\text{W}\,\text{K}^{-1}\,\text{m}^{-1}$ respectively $\lambda^{\text{G10-normal}}(4.2\,\text{K}) = 0.05\,\text{W}\,\text{K}^{-1}\,\text{m}^{-1}$ at 4.2 K to $\lambda^{\text{G10-warp}}(77\,\text{K}) = 0.41\,\text{W}\,\text{K}^{-1}\,\text{m}^{-1}$ respectively $\lambda^{\text{G10-normal}}(77\,\text{K}) = 0.32\,\text{W}\,\text{K}^{-1}\,\text{m}^{-1}$ at 77 K. Polyimide foil is an electrical and thermal insulator. Its thermal conductivity $\lambda^{\text{polyimide-foil}}$ increases monotonously with the temperature from $\lambda^{\text{polyimide-foil}}(4.2\,\text{K}) = 0.008\,\text{W}\,\text{K}^{-1}\,\text{m}^{-1}$ at 4.2 K to $\lambda^{\text{polyimde-foil}}(77\,\text{K}) = 0.14\,\text{W}\,\text{K}^{-1}\,\text{m}^{-1}$ at 77 K.

4.3.4.4 Filling materials

In HTS cables, the movement of the tapes has to be prevented in high force environments. The tapes have to be connected between themselves and in CICCs to the jacket to distribute mechanical loads evenly. This is achieved with filling materials. Filling materials should be mechanically strong, match the thermal expansion of *RE*BCO tapes and have high thermal conductivity to transport heat from the HTS tapes to the cooling system. There are two groups of possible filling materials: solders and glues or resins. Thermal and electrical conductivity are linked. Electrical conductors (e.g. solders) can be expected to exhibit much higher thermal conductivities than electrical insulators (e.g. glues and resins). As shown on the right of **figure 4.17**, in the temperature range from 4.2 K to 77 K the thermal conductivity of In50Sn50 solder increases from $\lambda^{\text{In50Sn50}}(4.2\,\text{K}) = 1.6\,\text{W}\,\text{K}^{-1}\,\text{m}^{-1}$ at 4.2 K to $\lambda^{\text{In50Sn50}}(77\,\text{K}) = 22.9\,\text{W}\,\text{K}^{-1}\,\text{m}^{-1}$ at 77 K. In the same temperature range, the thermal conductivity of Stycast blue remains below $\lambda^{\text{Stycast blue}} < 0.78\,\text{W}\,\text{K}^{-1}\,\text{m}^{-1}$. This means that the thermal conductivity of In50Sn50 solder is almost 30 times higher than that of Stycast blue. Pure Araldite epoxy resin exhibits an even lower thermal conductivity than Stycast blue. It is almost a thermal insulator. Mixing Araldite resin with quartz powder increases its thermal conductivity. At 45.5 % quartz powder content, the thermal conductivity of the mixture matches Stycast blue closely.

4.3.5 Summary and impact on HTS cables

To transport heat from the *RE*BCO tapes to the cooling system, high thermal conductivities of all materials are desired. The thermal conductivities of *RE*BCO tapes (**R**), structural materials (**S**), insulating materials (**I**) and filling materials (**F**) for HTS cables and their impact on cooling are summarized in **table 4.5**.

Table 4.5: Thermal conductivity (TC) λ of materials for high temperature superconductor cables: *RE*BCO tapes (**R**), structural materials (**S**), insulating materials (**I**) and filling materials (**F**).

	material	λ (4.2 K)	λ (77 K)	assessment
R	SCS12050 parallel to tape	$244\,\frac{\text{W}}{\text{K m}}$	$241\,\frac{\text{W}}{\text{K m}}$	very good thermal conductivity
	SCS12050 perpendicular to tape	$0.24\,\frac{\text{W}}{\text{K m}}$	$2.0\,\frac{\text{W}}{\text{K m}}$	low thermal conductivity, *RE*BCO layers are thermal shields
S	316-LN	$0.25\,\frac{\text{W}}{\text{K m}}$	$6.8\,\frac{\text{W}}{\text{K m}}$	reasonable thermal conductivity
I	G10 warp dir.	$0.08\,\frac{\text{W}}{\text{K m}}$	$0.41\,\frac{\text{W}}{\text{K m}}$	low TC, low fraction of material in magnets
	G10 normal dir.	$0.05\,\frac{\text{W}}{\text{K m}}$	$0.32\,\frac{\text{W}}{\text{K m}}$	low TC, low fraction of material in magnets
	polyimide foil	$0.008\,\frac{\text{W}}{\text{K m}}$	$0.14\,\frac{\text{W}}{\text{K m}}$	thermal insulator
F	In50Sn50 solder	$1.6\,\frac{\text{W}}{\text{K m}}$	$22.9\,\frac{\text{W}}{\text{K m}}$	good thermal conductivity
	Stycast blue	-	$0.78\,\frac{\text{W}}{\text{K m}}$	low thermal conductivity, complicates cooling
	Araldite	$0.05\,\frac{\text{W}}{\text{K m}}$	$0.13\,\frac{\text{W}}{\text{K m}}$	very low thermal conductivity
	Araldite + quartz powder	$0.08\,\frac{\text{W}}{\text{K m}}$	$0.86\,\frac{\text{W}}{\text{K m}}$	low thermal conductivity, complicates cooling

As shown in **subsection 4.3.4.1** and **subsection 4.3.4.2** the thermal conductivity of the *RE*BCO tapes itself is highly anisotropic. For stabilized tapes from SuperPower (SCS12050) [Supd] the thermal conductivity parallel to the tapes is almost 700 times higher than perpendicular to the tapes (at 22 K). For an HTS cable this means that the tapes shield each other thermally. This prevents heat from being transported through layers of *RE*BCO tapes to the cooling system. The impact of the highly anisotropic thermal conductivity of *RE*BCO tapes depends on the type of transposition in the cable.

- There are types of HTS cables which transpose the *RE*BCO tapes in layers. The layers thermally shield each other, reduce the radial thermal transport and lower the effective cooling contact area. This cannot be completely avoided with a central cooling channel, some layers of *RE*BCO tapes with bad connections to the cooling system will remain.

- There are types of HTS cables which transpose the *RE*BCO tapes, to repeatedly change positions from the center to the surface of the cable. There is no thermal shielding of the *RE*BCO tapes. Due to the high thermal conductivity along the cable (parallel direction), heat can be removed from all tapes efficiently. This types of HTS cables can be expected to exhibit superior cooling contact areas.

The thermal conductivities of structural and insulating materials are superior to *RE*BCO tapes in the perpendicular direction. Austenitic stainless steels such as 316-LN exhibit a much higher thermal conductivity. Insulating materials such as glass-fiber-reinforced-plastic G10, exhibit thermal conductivities comparable to *RE*BCO tapes in the perpendicular orientation. As the fraction of insulating materials in cables for fusion magnets is expected to be much lower than of the superconductors [JKS09], the impact of these insulating materials on the overall thermal conductivity will be negligible.

Filling voids and connecting the tapes to distribute the mechanical loads evenly may be possible with solders, glues and resins. As solders are electrically conductive, the thermal conductivity of solders is much higher than glues or resins. As shown on the examples of In50Sn50 solder and Stycast blue the thermal conductivity of the solder is up to 30 times higher than of that glue. Araldite epoxy resin is of even lower thermal conductivity than Stycast blue. However, mixing the resin with quartz powder increases its thermal conductivity. At a quartz powder content of 45.5 % it behaves very similarly to Stycast blue. In the frame of the efficient cooling of HTS cables, solders are preferred as filling materials compared with glues and resins.

5 Roebel Assembled Coated Conductor (RACC) cables

In some applications, such as fusion magnets[LMB09, SBOG10, LSI11], large superconducting generators [OLP09] and motors [KGSI06] or superconducting transformers [GSB11], the current carried by a single HTS tape is not enough. High current superconducting cables, consisting of many parallel tapes, are necessary. A concept for these cables is the Roebel assembled coated conductor (RACC) cable. Optimizations of the structure of these type of cables and an improved concept for contacting are shown in the following.

5.1 Introduction

At the beginning of the 20th century more and more powerful steam turbines were developed. The conversion to electrical power was challenging as the size of generators was limited. At that time several parallel copper rods were used for the conductor in the windings of generators. In such arrangements large areas are penetrated by magnetic fields inducing high coupling currents. The coupling currents result in significant heat loads limiting the size of the generators. Concepts with large conductor cross sections were not able to pave the way to larger generators.

In 1912 Ludwig Roebel found a solution. He proposed a conductor with two or more groups of insulated copper rods. These rods are interweaved with a braiding pattern with short transposition lengths as shown in a schematic drawing in **figure 5.1**.

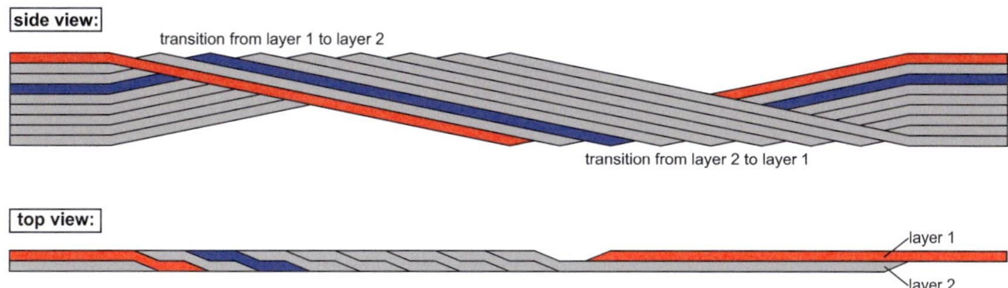

Figure 5.1: Schematic drawing of one half of a Roebelstab. The upper drawing shows the side view and the lower drawing the top view of a Roebelstab. Two conductors are emphasized in red and blue.

In this setup, several small areas are formed which are penetrated by magnetic fields instead of a single large one. The coupling currents and the heat loads are reduced significantly. This concept of multi-conductor cables was patented as the "Roebelstab" on march 19th 1912 [Sta12]. Using Roebelstaebe the construction of much larger, efficient generators became possible [ABB06]. Today, Roebelstaebe are still used in electrical applications.

5.2 High temperature superconductor Roebel cables

To use high temperature superconductors (HTS) in applications, the same basic principles apply as for the resistive generators in 1912. Complete transposition is necessary to minimize the coupling currents and the AC losses. Because of this, the assembling method for Roebelstaebe can be used with HTS tapes as proposed in 1999 [Wil99]. In 2006, this was first demonstrated with *RE*BCO tapes at the Forschungszentrum Karlsruhe GmbH (now the Karlsruhe Institute of Technology, KIT) carrying 500 A at 77 K in self-field [GNK06]. These second generation HTS Roebel cables are commonly referred to as Roebel Assembled Coated Conductor (RACC) cables.

The complete transposition of all tapes homogenizes the electrical and mechanical conditions of the tapes [TJS11] and reduces the coupling lengths. RACC cables have been shown to exhibit significantly lower alternating current (AC) losses compared with stacks of *RE*BCO tapes [SGK07, FHG08]. Furthermore, an HTS Roebel cable remains flexible and can be bended or twisted without degrading its current carrying capabilities. For a 16 strand RACC cable, the minimal bending angle is 11 mm and the minimum torsion angle is $20°\text{cm}^{-1}$ [GFH07], this allows the storage of RACC cables on cable rolls [Gol11] to simplify the construction of large devices.

5.2.1 Manufacturing of HTS Roebel cables

RACC cables are made of meander-shaped *RE*BCO tapes. At present the meander-shape is achieved by punching out trapezoids in regular intervals from both borders of a *RE*BCO tape. As shown in the upper drawing of **figure 5.2**, only half of the width of the *RE*BCO tape is used, the other half is wasted. The efficiency of the punching of the meander-structure can be increased by using significantly wider *RE*BCO tapes. Several meander-structured conductors are punched out of a wide *RE*BCO tape and this reduces the material loss to a small fraction of the width of the tape as shown in the lower drawing of **figure 5.2**. The increased material retention is essential for large scale applications and is today used in commercially available RACC cables [Lon11].

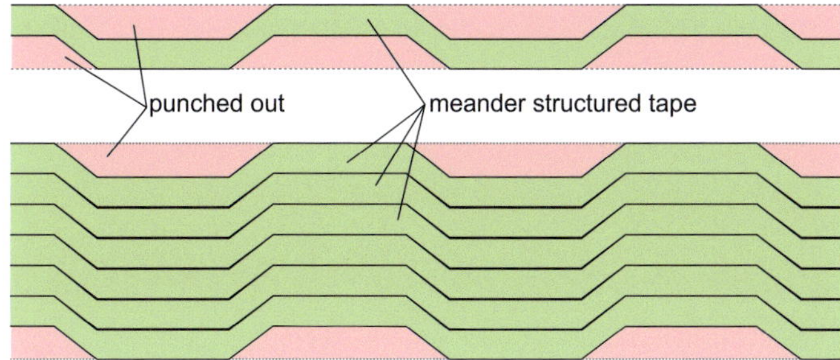

Figure 5.2: Punching methods for meander structured *RE*BCO tapes used for RACC cables. More than half of the width is lost if a single meander structure conductor is punched out of a *RE*BCO tape (upper picture). Punching several meander structured conductors form a much wider *RE*BCO tape is much more efficient. Only a fraction of the width of the tape is lost due to the punching (lower picture).

It has been shown that automated and reliable punching of the *RE*BCO tapes is possible with very high precision [Gol11, Lon11]. A high punching accuracy can be achieved with variations of less than 1 % of the tape width [GFK09]. If copper stabilized *RE*BCO tapes are used, the copper is partly smeared over the punched edge. This improves the mechanical contact to the copper layer and suppresses delamination of that layer. For best quality *RE*BCO tapes 3 % degradation of the current carrying capabilities occur due to punching [GFK09].

After punching meander structures into *RE*BCO tapes, the conductors are assembled using the Roebel assembling method. The assembly can be performed by hand [Gol11] or fully automated [Lon11]. Such an assembly is shown in a schematic drawing in **figure 5.3**.

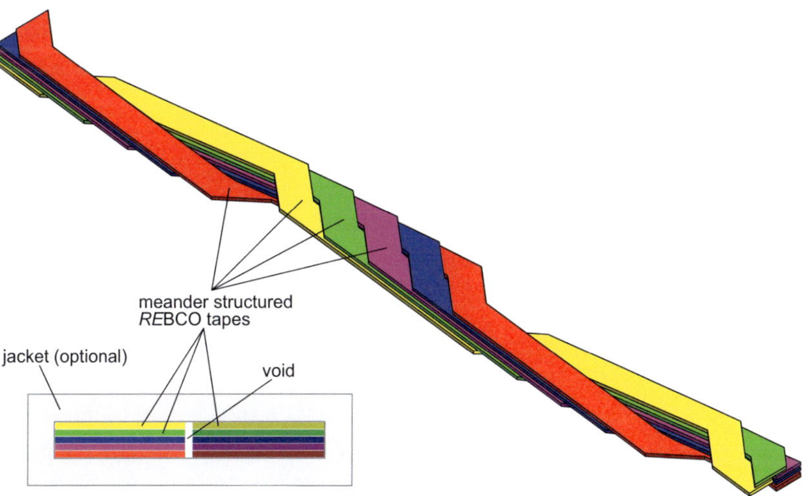

Figure 5.3: Schematic drawing of a part of a Roebel assembled coated conductor (RACC) cable. For better visibility only 5 tapes are shown. Picture after [SGK07].

This figure shows only half of a RACC cable. In a fully populated RACC cable there are only small voids between the tapes, a fully populated RACC cable is of bar geometry with a nearly planar surface, as shown in **figure 5.4**.

Figure 5.4: Roebel Assembled Coated Conductor (RACC) cable consisting of 10 strands. Picture from A. Kario.

5.3 Optimizing the geometry of the meander structure

*REB*CO tapes are resistant to tensile load by themselves, their critical stress limits are in the 500 - 700 MPa range [Haz10, Bru12, Supd, Sel11b]. The punched out sections of the meander structures required for RACC cables weaken the mechanical stability of the tapes. The inner corners or small radii of the punched out sections result in a sudden change in the geometric form of the tapes. This leads to stress concentrations, and locally increases the von-Mises stresses [Sen]. Inner corner or small radii are therefore the predetermined breaking points in meander structured tapes. The reduction in rigidity of the structured tapes depends on the abruptness of the change in shape [Sen]. This is determined by the geometry of the meander structure [Pil07]. However, the geometry also influences the space required for each crossing of the tapes. For a given transposition length the maximal number of tapes in a RACC cable is determined by the space required per crossing. In general, gradual transitions require more space per crossing therefore reducing the number of tapes in the cable.

Many superconducting applications such as fusion magnets or the rotors of motors and generators require high current while applying high tensile loads to their conductors at the same time. In these applications mechanical properties and the space required per crossing have to be balanced against each other. Determining the optimal geometry of the meander structures requires an in depth investigation of the influence of geometric parameters on mechanical properties and the space per crossing. In **section 5.3.1**, the geometry of meander structures of RACC cables is parametrized. The influence of the geometric parameters on the stress concentration and the space required per crossing is analyzed in **section 5.3.2**. The investigation is done by three dimensional (3D) simulations of single meander structured tapes with varied geometric parameters. For a given applied load, the von-Mises stress distribution is calculated and the maximum values compared. For each set of parameters, the length required for each crossing also is determined. Additionally, in any high current cable consisting of more than one tape, each tape is not only exposed to its self-field but to the combined field of all the tapes. As the tapes shield each other from the combined magnetic field [LPLL06], the current carrying capabilities of a tape are determined by its position in the multi-tape cable [VGT11]. Due to the complete transposition of a RACC cable, each strand of the Roebel assembly is in each position in one twist pitch. Therefore the degradation of the current carrying capabilities caused by the magnetic field varies along the length of each strand. On average it is the same for all strands on a long length scale ($\gg$ transposition length). In **section 5.3.3**, electro-magnetic analysis is performed. The magnetic field distribution of the 2D cross section of a RACC cable has been simulated by [VGT11] with finite element method (FEM) and the relative critical currents of each tape are calculated. This enables conclusions to be drawn about the current carrying capabilities along the length of the meander structure of each tape.

The results of the mechanical and electro-magnetic investigations are compared in **section 5.3.4**. A meander structure geometry is proposed which increases the tolerance of tensile loads and improves the current carrying capabilities. The investigation is published by the author in [BWVS12].

5.3.1 Parametrization of the meander structure

The meander structure required for Roebel assembled high coated conductor cables can be described by four geometric parameters. In **figure 5.5** a single section of the meander structure used for RACC cables is shown.

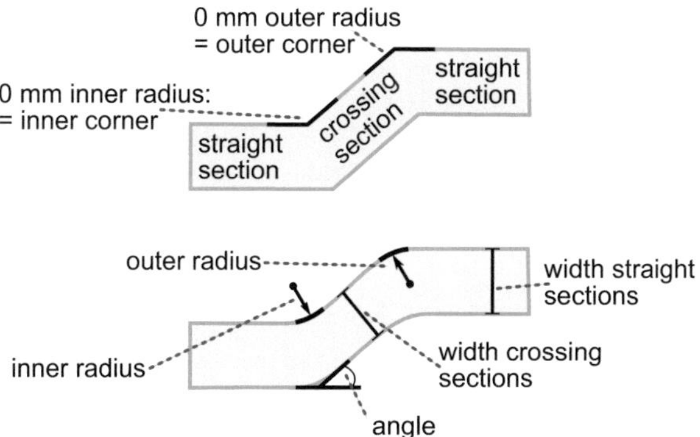

Figure 5.5: Single section of the meander structure used for Roebel assembly. Straight sections and crossing section of the meander structure with intermediate sharp transitions. Sharp transition on the outside are referred to as "outer corners" or as an outer radius r_{out} of 0 mm. Sharp transitions on the inside are "inner corners" or as an inner radius r_{in} of 0 mm (upper picture). Geometric parameters of the meander structure: angle θ, inner radius r_{in}, outer radius r_{out}, width of the crossing sections $w_{crossing}$ and width of the straight sections $w_{straight}$ (lower picture).

The "straight sections" and the "crossing section" are shown in the upper picture. If the transitions between these sections are sharp they are referred to as "inner corners" or "outer corners". Smooth transitions between the straight sections and the crossing sections are described by their radii. There are inner radii r_{in} and outer radii r_{out}. In plots, inner corners and outer corners are also referred to as inner radii r_{in} and outer radii r_{out} of 0 mm, respectively. In the lower part of **figure 5.5** the geometric parameters of the meander structure are specified.

The parameters are the inner radius r_{in}, the outer radius r_{out}, the angle θ and the relative width w_{rel}, which is the quotient of the width of the crossing sections $w_{crossing}$ and the width of the straight sections $w_{straight}$ as shown in **equation 5.1**.

$$w_{rel} = \frac{w_{crossing}}{w_{straight}} \qquad\qquad [5.1]$$

Outer radius r_{out} and inner radius r_{in} describe the radius of the corresponding round transitions between the crossing sections and the straight section of the meander structure. The angle θ is the angle between the direction of the crossing sections and the straight sections.

All following investigations will use this terminology to describe the meander structure of Roebel Assembled Coated Conductor (RACC) cables.

5.3.2 Influence of geometric parameters on mechanical properties

In this section the influence of the geometric parameters on the stress concentration and the space required per crossing is investigated.

5.3.2.1 Simulation method

Three dimensional models of 4 mm and 12 mm wide *RE*BCO tapes with two meander structures are created with varying geometric parameters. The mechanical properties of *RE*BCO tapes manufactured by SuperPower (data from 2010) are used for the models. All tapes are modeled as a monolithic material with a thickness of 100 µm [Haz10, page 2] and a Young's modulus of 120 GPa [Haz10, page 18]. A Poisson's ratio of 0.311 is used. This ratio is calculated as the weighted average of the Poisson's ratios of the Hastelloy substrate 0.32 [Hay] and the copper stabilization 0.33 [Wikc]. Constant tensile loads of 1 kN are applied on one side of the modeled tapes while the other sides are fixed. On the straight sections and the tape plane, roller constraints are used to allow movement and deformation along the direction of the tape. The numerical calculations are performed with the structural mechanics module of the commercial finite element method (FEM) simulation software package Comsol. The distribution of the von-Mises stress is calculated and visualized as a surface plot of the meander structure as shown in **figure 5.6**.

The maximum value of the von-Mises stress distribution is used as a measure of the sensitivity of the corresponding geometry to tensile loads. Higher von-Mises stress means that the stress is

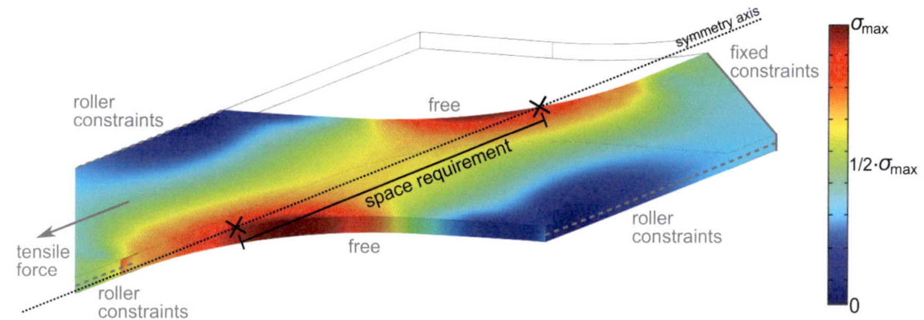

Figure 5.6: Crossing section of a meander structured tape with applied tensile strain. Constraints and simulation result of a 4 mm wide tape with an angle of 30° and an inner radius of 8 mm. The local stress of a region is shown as a gradient from low stress (blue) to high stress (red). The deformation of the meander structure is strongly exaggerated to improve visibility.

concentrated in a small area. High stress areas favor the growth of cracks and mechanical failure which results in a mechanically unstable RACC cable. If the maximum von-Mises stress is low, the stress is distributed over a large area and such a cable can withstand higher tensile loads.

For each set of parameters, the distance between the intersection points of the borders of the crossing section with the symmetry axis is calculated as shown in **figure 5.6**. This distance is a measure for the length required for the crossing of these meander structures and is thus referred to as the "space requirement" s_{req}. For a given number of tapes, it determines the minimal possible transposition length (twist pitch) of the respective RACC cable. For high current Roebel cables consisting of several hundreds of tapes the distance required for each crossing becomes an important factor. As an example, a RACC cable able to carry currents larger than 10 kA needs an effective *RE*BCO conductor width of approximately 2 m [Gol11, page 28]. This means 200 or more tapes which results in large transposition lengths, even with the multi-stacking of tapes. If the transposition length is too large, high coupling currents [Iwa09, page 406 et sqq.] and high AC losses occur [Kom95, p. 64 et sqq.][Wil83, p. 174 et sqq.] which limit the applicability of the corresponding cable design. Therefore for a high current RACC cable each crossing should require as little space as possible (have a low "space requirement" s_{req}) to fit several hundred tapes within a reasonable transposition length.

5.3.2.2 Simulation results

These simulations are run with different sets of geometric parameters. The results shown are for a 4 mm wide, copper stabilized, coated conductor from SuperPower (SCS4050). The default parameter set is shown in **table 5.1**. The default configuration is chosen as the reference because RACC cables have already been built and tested with that specific set of geometric parameters [GFK09]. For comparison, the relative maximal von-Mises stress is given in the calculations which is the quotient of the maximal von-Mises stress of the used set of parameters and the default configuration.

Table 5.1: Default geometry parameters of Roebel cables (as used in a 2.6 kA Roebel cable [GFK09])

angle θ	inner radius r_{in}	outer radius r_{out}	relative width w_{rel}
30°	0 mm $\Rightarrow$ inner corner	0 mm $\Rightarrow$ outer corner	1

All results are grouped into three graphs in which one or two parameters are varied while the others are kept constant.

In **figure 5.7** the dependency of the maximal occurring von-Mises stress (y-axis) on inner radii r_{in} (x-axis) is determined for specific outer radii r_{out} (shown in the legend). A large inner radius distributes the mechanical stress over a large area and avoids stress concentration [Sen]. Large inner radii r_{in} are therefore essential for mechanically solid RACC cable. Outer radii r_{out} increases the maximum occurring von-Mises stress, which is highly disadvantageous. Thus, RACC cables always must be designed with outer corners ($r_{out} = 0$ mm) in their meander structures. All following calculations use outer corners.

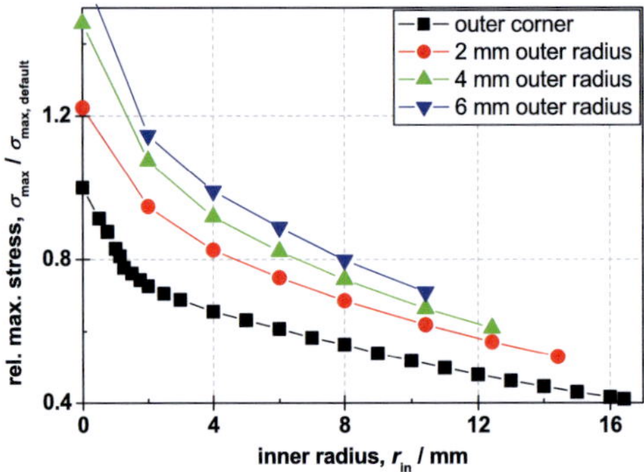

Figure 5.7: Maximal occurring von-Mises stress for different inner radii r_{in} and outer radii r_{out}. An angle θ of 30° and a relative width w_{rel} of 1 are assumed.

In **figure 5.8** the dependency of the maximal occurring von-Mises stress (y-axis) on the angle θ (x-axis) is determined for specific inner radii r_{in}. Small inner radii r_{in} and especially inner corners ($r_{in} = 0$ mm) result in high von-Mises stresses regardless of the angle θ used. With increasing inner radii r_{in} the angles θ become more and more important. The relative maximal von-Mises stresses form curves with peaks at specific angles. These peaks must be avoided in a mechanically solid cable. Thus inner radii r_{in} of 6 mm and larger require angles θ of 30° and more.

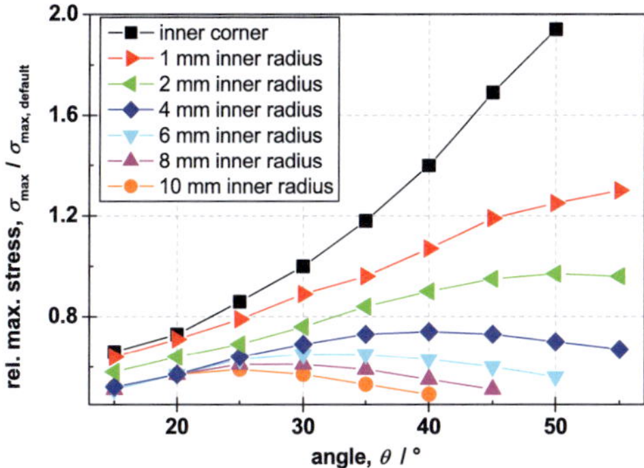

Figure 5.8: Dependency of maximal occurring von-Mises stress on angle θ for different inner radii r_{in}. An outer corner ($r_{\text{out}} = 0\,\text{mm}$) and a relative width w_{rel} of 1 are assumed.

Figure 5.9 shows the influence of the parameter relative width w_{rel} (x-axis) on the stress distribution (y-axis) for specific sets of angles θ and inner radii r_{in} (shown in the legend). Thin crossing sections with relative widths $w_{\text{rel}} < 1$ result in high von-Mises stresses and mechanically weakened meander structures. They decrease the area onto which the load is distributed. For a mechanically solid cable, the crossing sections should be wider than the straight sections; relative widths w_{rel} larger than 1 are necessary. At the same time, relative widths w_{rel} larger than 1.2 are also not recommended as further increases in width w_{crossing} of very wide crossing sections reduces the occurring stresses insignificantly. Such extremely wide crossing sections ($w_{\text{rel}} > 1.2$) only increase the "space requirement" s_{req}, meaning they require more space along the cable for each crossing and reduce the number of tapes for a given twist pitch.

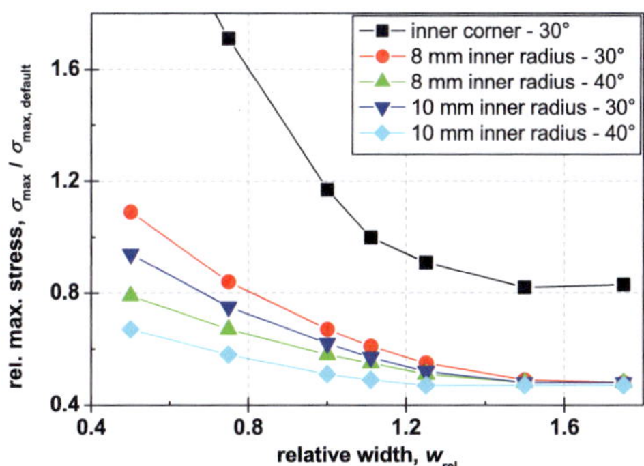

Figure 5.9: Dependency of maximal occurring von-Mises stress on the relative width of the crossing sections w_{rel} for different sets of inner radii r_{in} and angles θ. Outer corners ($r_{\text{out}} = 0\,\text{mm}$) are assumed.

The contribution of the parameter relative width w_{rel} to the "space requirement" s_{req} can be calculated by dividing the width of the crossing section w_{crossing} by the sine of the angle θ as shown in **equation 5.2**.

$$s_{\text{req}}^{\text{width}} = \frac{w_{\text{crossing}}}{\sin(\theta)} \qquad [5.2]$$

The contribution of the other geometric parameters to the "space requirement" s_{req} is shown in **figure 5.10**. Similar to the von-Mises stress distribution, the "space requirement" s_{req} is given relative to the requirement of the default configuration. The y-axis shows the normalized "space requirement" s_{req} depending on the angle θ (x-axis) for specific inner radii r_{in} (shown in the legend). In general, higher inner radii r_{in} require more space, hence a mechanically solid Roebel geometry requires more space per crossing and has a larger minimal transposition length for a given number of tapes.

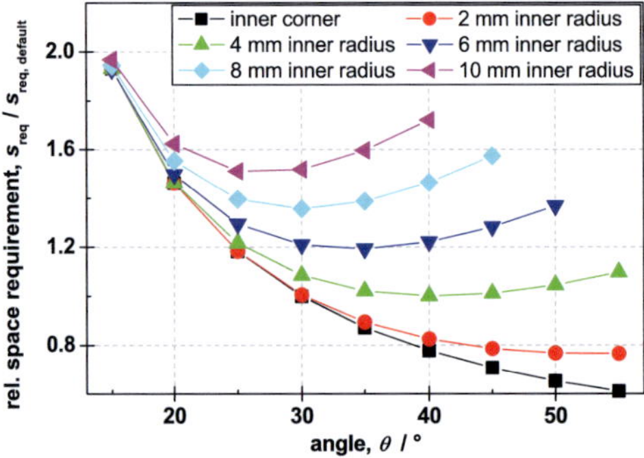

Figure 5.10: Required space per crossing ("space requirement" s_{req}) for different sets of inner radii r_{in} and angles θ. An outer corner ($r_{\text{out}} = 0\,$mm) and a relative width w_{rel} of 1 are assumed.

5.3.2.3 Impact on Roebel geometry

As mechanical stability and minimal transposition length are influenced by the geometry parameters of the meander structure used in RACC cables, stability and minimal transposition length must be balanced against each other while choosing an optimal set of geometric parameters. The crossing sections should always be slightly wider than the straight sections (relative width w_{rel} of 1.05 to 1.2) because thin crossing sections strongly degrade the mechanical stability of the meander structure.

The meander structure can be optimized for low space per crossing, to assemble Roebel cables consisting of several hundreds of tapes with reasonable transposition lengths. In such cases, outer corners ($r_{\text{out}} = 0\,$mm), inner radii r_{in} of 6 mm and angles θ of 35° are good choices to deliver good mechanical stability. If the minimal transposition length is not the dominant factor (for example in RACC cables consisting of only a few tapes), outer corners ($r_{\text{out}} = 0\,$mm), inner radii r_{in} larger than 8 mm and angles θ greater than 35° are preferred sets of parameters for mechanically strong cables.

5.3.3 Influence of geometric parameters on current density

In this section the influence of the geometric parameter relative width w_{rel} on the current carrying capabilities of a RACC cable is investigated. As discussed in **section 5.3.1**, the relative width w_{rel} is the factor by which the crossing sections are wider than the straight sections. An optimal relative width w_{rel} is obtained by simulating the distribution of the magnetic flux density in a two dimensional cross section view of a 14 tape RACC cable. The simulations are performed in two steps. First a single tape is modeled and compared with field dependent critical current measurements. The validated model and the parameters are then used to simulate a RACC cable. This method has been proposed by [VGT11].

5.3.3.1 Simulation method

The electro-magnetic FEM calculations assume infinitely long straight conductors with current flowing only perpendicular to their cross sections. The *REBCO* tape is modeled in real dimensions. The tape is represented by its superconducting layer with a thickness of $1.4\,\mu\text{m}$. All other components such as substrate, buffer and stabilization are modeled as vacuum because their influences to the magnetic flux or the current are negligible.

The numerical calculations are done with the commercial FEM program Comsol in a two dimensional magneto static model. The state variable in the model is the vector potential $\vec{A}$. Bean's model of the critical state [Bea62] is used to describe the superconductor's properties. The critical current of a tape is reached when the complete cross-section of the tape is filled by its critical current density, dependent on the local magnetic field. The critical current of a tape is calculated by integration of the critical current density over the strand's cross-section.

For the description of the $J_{\text{C}}\left(\vec{B}\right)$ dependence, the functions proposed by Pardo et al. [PVGv11] are used.

$$J_{\text{C}}\left(\vec{B}\right) = \left(J_{\text{C}_{\text{ab}}}^{m} + J_{\text{C}_{\text{c}}}^{m} + J_{\text{C}_{\text{i}}}^{m}\right)^{\frac{1}{m}} \qquad [5.3]$$

In **equation 5.3**, m is constant and $J_{\text{C}_{\text{ab}}}$, $J_{\text{C}_{\text{c}}}$ and $J_{\text{C}_{\text{i}}}$ are defined by **equation 5.4**, **equation 5.5** and **equation 5.6**.

$$J_{\text{C}_{\text{ab}}} = \frac{J_{0_{\text{p}}}}{\left(1 + \frac{B_{\text{loc}} \cdot f_{\text{ab}}}{B_{0_{\text{ab}}}}\right)^{\beta}} \qquad [5.4]$$

$$J_{C_c} = \frac{J_{0_p}}{\left(1 + \frac{B_{\mathrm{loc}} \cdot f_c}{B_{0_c}}\right)^{\beta}} \tag{5.5}$$

$$J_{C_i} = \frac{J_{0_i}}{\left(1 + \frac{B_{\mathrm{loc}} \cdot f_i}{B_{0_i}}\right)^{\alpha}} \tag{5.6}$$

The dependence on the local magnetic field B_{loc} and its direction θ_{loc} is included in the functions f_{ab}, f_c and f_i as described by **equation 5.7** to **equation 5.9**.

$$f_{\mathrm{ab}} = \begin{cases} \sqrt{\cos^2\left(\theta_{\mathrm{loc}} - \delta_{\mathrm{ab}}\right) + u_{\mathrm{ab}}^2 \sin^2\left(\theta_{\mathrm{loc}} - \delta_{\mathrm{ab}}\right)} & \text{if } \theta_{\mathrm{loc}} \in \left[-90° + \delta_{\mathrm{ab}}, 90° + \delta_{\mathrm{ab}}\right] \\ \sqrt{v^2 \cos^2\left(\theta_{\mathrm{loc}} - \delta_{\mathrm{ab}}\right) + u_{\mathrm{ab}}^2 \sin^2\left(\theta_{\mathrm{loc}} - \delta_{\mathrm{ab}}\right)} & \text{otherwise} \end{cases} \tag{5.7}$$

$$f_c = \sqrt{u_c^2 \cos^2\left(\theta_{\mathrm{loc}} - \delta_c\right) + \sin^2\left(\theta_{\mathrm{loc}} - \delta_c\right)} \tag{5.8}$$

$$f_i = \sqrt{\cos^2\left(\theta_{\mathrm{loc}} - \delta_c\right) + u_i^2 \sin^2\left(\theta_{\mathrm{loc}} - \delta_c\right)} \tag{5.9}$$

An external magnetic field can be included by adjusting the boundary conditions with the corresponding vector potential $\vec{A}_{\mathrm{B}}$. The vector potential $\vec{A}_{\mathrm{B}}$ is shown in **equation 5.10** for an external field of the strength H_{ext} and angle θ_{ext}.

$$\vec{A}_{\mathrm{B}} = \mu_0 \cdot H_{\mathrm{ext}} \cdot \left(y \cdot \cos\left(\theta_{\mathrm{ext}}\right) - x \cdot \sin\left(\theta_{\mathrm{ext}}\right)\right) \tag{5.10}$$

In a first step, a single 4 mm SuperPower *REBCO* tape is simulated and the dependency of the critical current density on the strength and angle of a magnetic background field is calculated. The calculated data is compared with in-field critical current measurements [SR] of the exact type of HTS tape. The measured and the calculated critical current density dependencies are compared in **figure 5.11**. Measured data is shown as points while the results of the FEM models are lines. Measured and calculated data are almost a perfect fit. They are in good agreement with data published by the tape manufacturer [SXM09], too. Because of this, the simulation is suitable for describing the superconducting properties of *REBCO* tapes in detail. Values of the parameters of **equation 5.3** to **equation 5.9** of these simulations are given in **table 5.2**.

In a second step a RACC cable with 14 strands is modeled and simulated. The RACC cable is described by its cross-section view in two dimensions. The same simulation method and identical simulation parameters are used as for the single tape simulations. Therefore, the superconducting properties of the *REBCO* material will be described accurately. The simulation is done with self-field conditions, so no additional vector potential is needed. The RACC cable is described as 14 strands in 2 columns of 7 tapes each. The columns are horizontally separated by 0.4 mm of vacuum. The calculations assume infinitely long straight conductors with current flowing

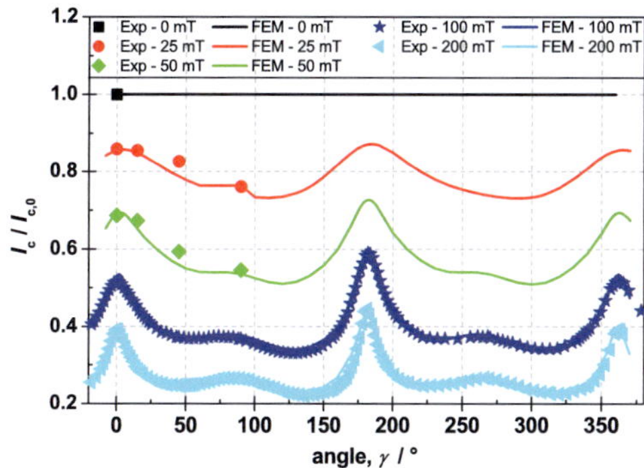

Figure 5.11: Dependency of the relative critical current density (y-axis) of 4 mm SuperPower *REBCO* tape (from 2010) on strength (listed in the legend) and angle (x-axis) of a magnetic background field. The points are measured values while the lines are simulated. Data from [SR, VGT11].

Table 5.2: Parameters used in the electro-magnetic FEM simulations. Parameters are for 4 mm wide SuperPower *REBCO* tape manufactured in 2010.

smoothing		*field dependence*		*angular dependence*	
parameter	value	parameter	value	parameter	value
m	8	J_{0_p}	$53\,\mathrm{GA\,m^{-2}}$	δ_{ab}	$1.7°$
		J_{0_i}	$32.12\,\mathrm{GA\,m^{-2}}$	δ_c	$-7.2°$
		$B_{0_{ab}}$	$4.569\,\mathrm{mT}$	u_{ab}	8.337
		B_{0_c}	$2.016\,\mathrm{mT}$	u_c	1.794
		B_{0_i}	$32\,\mathrm{mT}$	u_i	1.7
		β	0.4764	v	0.9
		α	0.9		

only perpendicular to their cross sections. The tapes are represented by their superconducting layers each with a thickness of $1.4\,\mu\mathrm{m}$ and a width of $1.98\,\mathrm{mm}$. The other components such as substrate, buffer, stabilization or air between the tapes are modeled as vacuum. This results in an vertical separation of the superconducting layers of $147\,\mu\mathrm{m}$. This configuration is shown schematically in **figure 5.12**.

5.3.3.2 Simulation results

The distribution of the magnetic flux density in the cross section view of the RACC cable has been calculated by [VGT11] and is shown in **figure 5.13**. The superconducting layers shield each other from the magnetic flux [LPLL06, GAS06]; therefore the flux density is lowest in the center regions and highest near the border of the RACC cable.

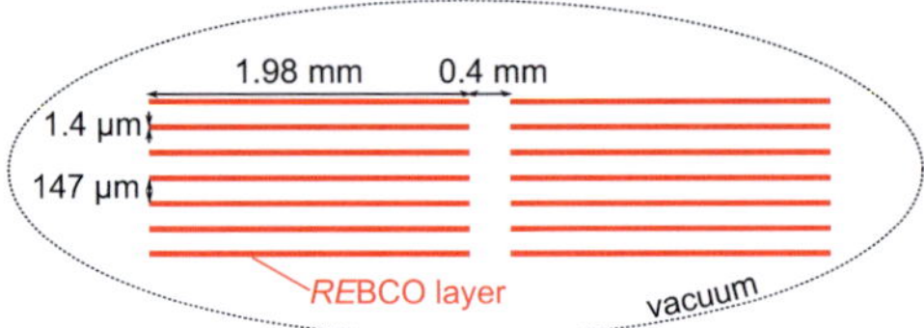

Figure 5.12: Schematic drawing of the 2D model of RACC cables used for electro-magnetic simulations. The RACC cable is modeled as 14 superconducting layers in two columns separated by vacuum. Drawing is not to scale.

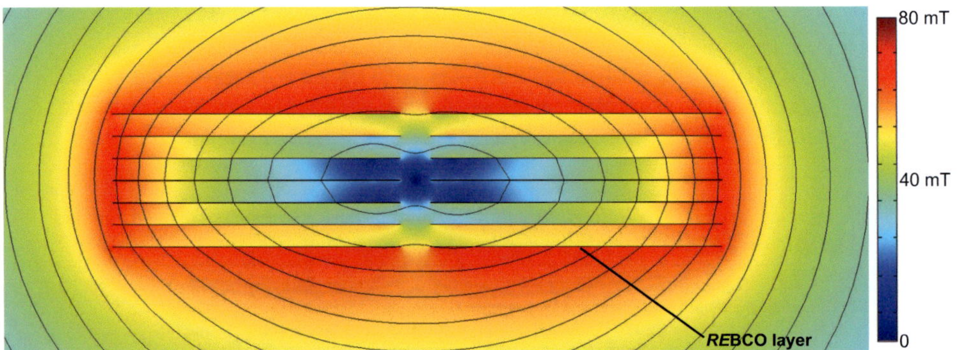

Figure 5.13: Simulated magnetic flux density distribution of RACC cables consisting of 14 tapes. The distribution of the magnetic flux density is shown as transition from low flux density (blue) to high flux density (red). Drawing is true to scale. Due to the very low thickness (1.4 µm) of the superconducting layers they are only visible as horizontal black lines. Picture and data from [VGT11].

Using the measured correlation between the critical current density, angle and value of the magnetic field (see **figure 5.11**), the critical current density distribution in each superconducting layer can be calculated. The higher magnetic field in the border regions reduces the critical current density in these tapes while the center tapes are shielded from the field and retain a high critical current density. The normalized current density $j_{c,norm}$ of a tape is obtained by integrating the local critical current density of the corresponding tape and dividing by the average current density of the center tapes. As shown in **figure 5.11**, the dependency of the critical current density on the angle of a magnetic background fields is not symmetrical (neither to $0°$ nor $90°$). The anisotropy of the SuperPower *REBCO* is shifted to one side. Therefore the degradation of the current carrying capabilities is different for the left and the right stack of superconducting layers. This is shown for each of the 14 strands of the investigated cable in **figure 5.14**.

Due to the complete transposition of a RACC cable, the magnetic and electric boundary conditions are the same for each tape on a long length scale. During one twist pitch, each tape has been in each position.

position	$j_{c,norm}$	position	$j_{c,norm}$
11/14	0.81	10/14	0.76
12/14	0.87	9/14	0.82
13/14	0.95	8/14	0.90
0 & 1	1.00	7/14	1.00
1/14	0.90	6/14	0.94
2/14	0.81	5/14	0.86
3/14	0.75	4/14	0.80

Figure 5.14: Normalized critical current density $j_{c,norm}$ of each tape (floating point numbers) of a 14 strand RACC cable and the corresponding positions (fractional numbers) along the meander structure for a single tape. Data from [VGT11, BWVS12].

This means that the cross section view of the cable can be represented by different positions along the length of a single tape. As shown in **figure 5.15** for the 14 strands RACC cable, there are 14 positions along the meander structure of a tape which are the equivalent to the strands in the cross section view (**figure 5.14**). This transforms the two dimensional critical current distribution (**figure 5.14**) to one dimension as shown in **figure 5.16**. The average current carrying capabilities of the different tapes are embodied by the critical current densities in specific positions along the length of a meander structured tape. The dots are the data points gained from the two dimensions (2D) to one dimension (1D) transformation while the connecting line is a linear interpolation in each segment.

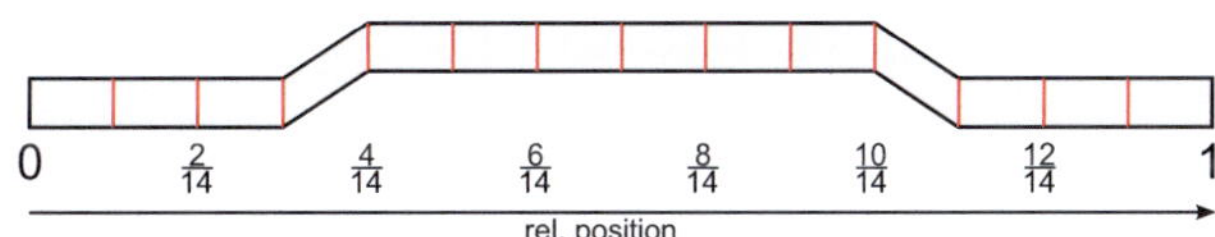

Figure 5.15: Corresponding positions in the meander structure of a single tape to the strands in the cross section view of a Roebel cable with 14 strands.

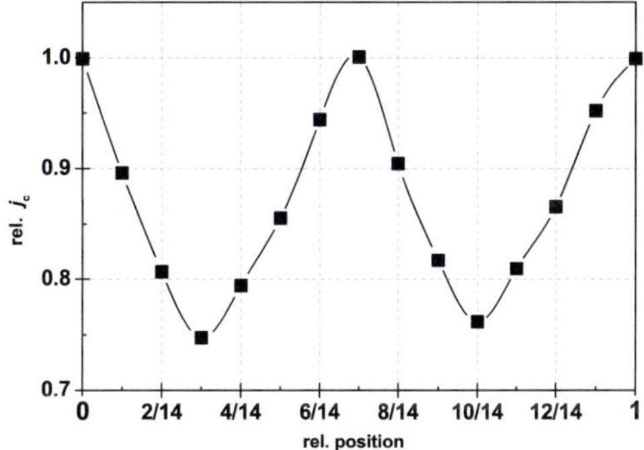

Figure 5.16: Relative critical current density along the meander structure of a tape for a RACC cable consisting of 14 tapes. Data from [VGT11, BWVS12].

5.3.3.3 Impact on Roebel geometry

In a RACC cable the "crossing sections" (as visualized in **figure 5.5**) are always on the outside of the cable (top and bottom) while the connecting "straight sections" go through the center. Therefore the "crossing sections" are the areas of each strand with the highest magnetic flux density and lowest current carrying capabilities. This can be seen by combining the numbers of the 14 positions from **figure 5.15** with the critical current density distribution of **figure 5.16**. The "crossing sections" (positions $\frac{3}{14}$ to $\frac{4}{14}$ and positions $\frac{10}{14}$ to $\frac{11}{14}$) are always on the outside and have the lowest critical current density. For the current carrying capabilities, they are the weak areas of the cable. The center of the "straight sections" (positions $0, 1$ and $\frac{7}{14}$) have a much higher critical current density as they are always on the inside. These areas are shielded from the magnetic field.

In a long RACC cable (with ideal contacts and homogeneous HTS material) all tapes carry the same average current [TJS11]. If the strands are isolated from each other, which means zero current sharing, the weakest areas of each strand determine the maximal current of that strand. In this case the lower critical current in the crossing sections dictates the current for each strand. As the cable is transposed each strand exhibits the same weak spots ("crossing sections") and good spots (when it is in the center of the RACC cable). This causes all the strands to carry the same current.

If the strands are connected without any resistance, balancing current can share the current between the strands. This is called complete current sharing. In this case the lower critical current densities in the "crossing sections" do not define the maximal current of each tape. Balancing currents allow the current to redistribute; this means that the reduction of the overall current due to the weak areas is less severe. The tapes still carry the same average current in one transposition length, but this current is higher than in the case of zero current sharing.

In a real RACC cable, the level of current sharing is between these two extremes. There is always some current sharing, but it is far from complete. The real level of current sharing can be determined by calculating the transport current of a Roebel cable from its simulated current density distribution for both case and compare these with the measured transport current of the same cable. This comparison is done in **table 5.3** [TVP10] for the investigated Roebel cable.

For the investigated RACC cable there is almost no current sharing, therefore the critical

Table 5.3: Simulated and measured current carrying capabilities of the 14 strand Roebel cable. Data from [TVP10].

measured I_c	calculated I_c *complete current sharing*	calculated I_c *no current sharing*
465 A	496.3 A	463.3 A

current densities in the crossing sections dictate the current of each strand. This reduction of the current carrying capabilities can be lowered by changing the geometry of the meander structure. For the tapes on the outside to carry the same current as the second tapes from the outside, their cross section area has to be increased. It has to be increased by the reciprocal of the factor by which their critical current density is lower. The top and bottom tapes are always the "crossing sections" of the meander structure therefore this can be done by increasing the geometric parameter relative width w_{rel}. Now they are not the weak spots any more. This does not change the outer dimensions of the cable and only slightly increases the minimal possible transposition length. This increase is calculated for the investigated RACC cable for top and bottom crossing sections in **table 5.4**. With no current sharing the "crossing sections" should be about 7 % wider than the straight sections to increase the overall current of this cable by almost the same amount.

Table 5.4: Required increase in width of the crossing sections to compensate for lower critical current density.

	min. rel. j_{c} *(crossing sections)*	max. rel. j_{c} *(second tapes from the outside)*	required increase in width w
top	76 %	97 %	10.4 %
bottom	75 %	86 %	10.7 %

5.3.4 Summary and recommendation

For designing any RACC cable, the geometry of the meander structures with its four parameters inner radius r_{in}, outer radius r_{out}, angle θ and relative width w_{rel} is an important aspect as it influences the mechanical properties, the minimal achievable transposition length and even the current carrying capabilities of the whole cable. In RACC cables the crossing sections should be at least to 10 % wider than the straight sections. This compensates for the reduction of the critical current density in the crossing section due to the high magnetic flux density near the border. Wider crossing sections increase the overall current of the RACC cable by up to 10 % depending on the level of current sharing. The optimal value of the parameter relative width w_{rel} has to be calculated for each RACC cable as it depends on the HTS material used, the number of strands and the current sharing. In the investigated case of a 14 strand Roebel cable, zero current sharing has shown to be a good approximation.

Additionally, wide crossing sections are also beneficial for mechanical stability while allowing almost the same transposition length. "Inner corners" must be avoided as they lead to areas in the stress is concentrated under applied tensile load. Such areas mechanically weaken a RACC cable, they are predetermined breaking points. Large inner radii r_{in} distribute the stresses to larger areas and therefore increase the mechanical stability. This must be balanced against an increase in the space required per crossing which increases the minimal possible transposition length for a given number of tapes. This can be a limiting factor, especially for high current RACC cables with several hundreds of tapes. Outer radii r_{out} must never be used as they mechanically weaken the cable and are bad for the twist pitch. As a summary, optimized sets of geometric parameters are given in **table 5.5** for two cases: "mechanical stability" and "high number of tapes".

Table 5.5: Optimized set of geometry parameters for the meander structure of RACC cables.

	angle θ	inner radius r_{in}	outer radius r_{out}	relative width w_{rel}
optimize for mechanical stability	$>35°$	$>8\,$mm	outer corner	>1.1
optimize for high number of tapes	$35°$	$6\,$mm	outer corner	>1.1

5.4 Optimizing the contacts of HTS Roebel cables

Good electrical contacts are essential for optimal coated conductor cable performance. The resistivity of the *REBCO* is negligible due to its superconducting properties. Thus, the overall resistivity of each tape is mainly determined by the resistivity of its contacts. In short cables or cables of direct current (DC) applications, homogenous contacts are necessary for all tapes to carry the same current.

5.4.1 State-of-the-art

Commonly used contacts of RACC cables consist of a U-shaped copper profile with an open area slightly wider than the width of the RACC cable. The profile has a length of one twist pitch of the corresponding RACC cable. The inside of the copper U-shape is coated with an indium-tin based solder with a low melting point (In50Sn50 solder which melts at 116 - 127 °C is often used). Uniform coating with solder is achieved by using a fluxing agent and by heating the whole copper U-shape above the melting point of the solder with a heating plate. One transposition length of the RACC cable is also coated with such a low temperature solder. *REBCO* tapes can easily be damaged through acids. Thus only acid free fluxing agents can be used to coat the tapes of the RACC cable [AMS12]. The solder coated end of the RACC cable is now inserted in the U-shaped copper structure. Pressure is applied while any voids are filled with additional solder [Lon11]. Such a contact is shown schematically in **figure 5.17**.

With this method, the contact resistances per tape are commonly in the $100 - 500\,$n$\Omega\,$cm^2 range [GFK09, Kud12]. The homogeneity however can be rather low. As shown in a schematic

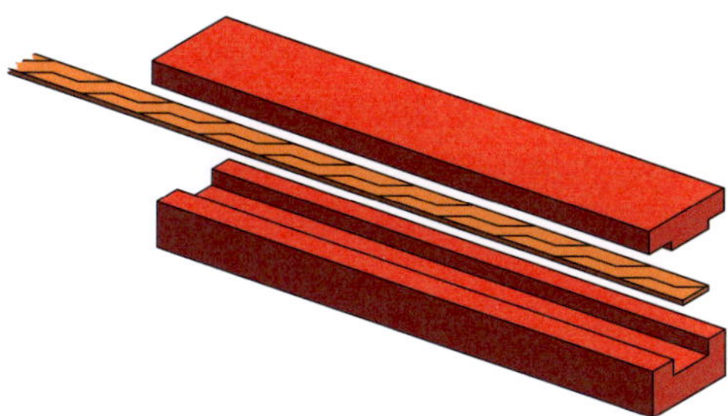

Figure 5.17: Schematic drawing of contacts of RACC cables. Copper current leads are shown in red. The RACC cable in brown.

cross section view in **figure 5.18**, only the bottom tapes are connected to the copper current lead over their whole width. All other tapes are only connected through their outer sides. Therefore such contacts of RACC cables must be of the length of one twist pitch of the cable allowing each tape to be on the bottom once. Even then, as the contact resistance to the outer side of the *RE*BCO tapes strongly varies, significant differences in the contact resistances of the individual tapes can occur.

A current through the RACC cable distributes according to the resistivity of the tapes. Therefore, inhomogeneous contact resistance means that not all tapes carry the same current. The tapes featuring a lower contact resistance will carry a larger fraction of the current. If the current of the cable is increased these low-resistance tapes will reach their critical electric field E_c much earlier. Tapes which are connected to the current lead with higher resistance remain at low electric fields up to very high currents. The tapes with low contact resistance therefore seem to limit the current carrying capabilities of the cable. A RACC cable with inhomogeneous contacts does not fully use the current carrying capabilities of all tapes. The critical electric field of the low-resistance tapes is reached without the high-resistance tapes carrying full current. Low homogeneity contacts therefore reduce the current carrying capabilities of the cable.

The electric field across the whole cable (measured as the voltage drop between the contacts) corresponds to a superconductor with a low n-value. This behavior is shown in **figure 5.19** in an

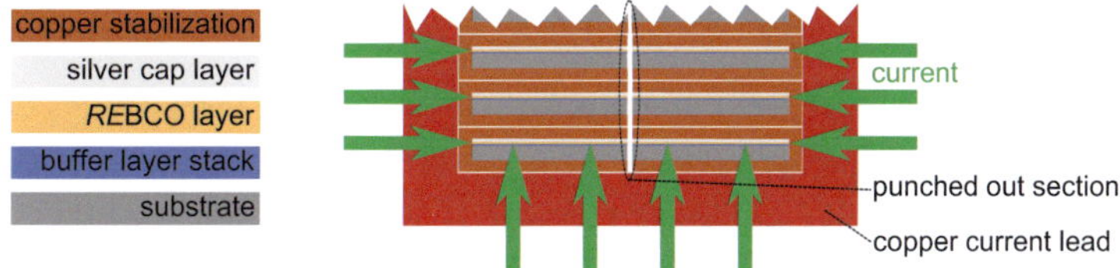

Figure 5.18: Schematic drawing of contacts of RACC cables in cross section view. The bottom tapes are connected over their whole width to the copper current lead; all other tapes only through their outer sides. Due to the punched out sections, the stacks of *RE*BCO tapes are not directly coupled. Picture after [GFK09].

electric field vs. current plot of a 45 tape RACC cable in a liquid nitrogen bath. The slope and the current at which the transition from the superconducting state to normal conduction occurs is very different for the individual tapes and the whole cable[a]. The slope of the transition is very gentle for the whole cable which corresponds to a low n-value (see **subsection 2.1.2.3**).

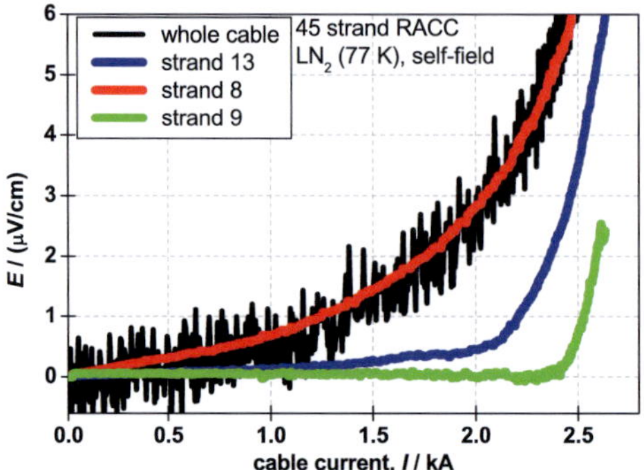

Figure 5.19: Electric field vs. current graph of a RACC cable measured in liquid nitrogen (77 K). The cable consists of 45 *REBCO* tapes from SuperPower and was assembled in 2008 [GFK09]. This is an extreme example for inhomogeneous contact resistance. The inhomogeneities of the contact resistance are very high as multi-stacking of tapes was used. Additionally, with a total length of 10 cm, the contacts are rather short. The electric field of the tapes (strand 8, strand 9 and strand 13) is measured with nano volt meters while the electric field over the contacts, referred to as "whole cable" is measured with the quench protection system. Thus the whole cable shows significant more noise compared with the individual tapes.

With this method of contacting, n-values of 3 to 9 are common in RACC cables [Lon12]. To further improve the n-value of RACC cables, contacts for all the tapes with nearly identical contact resistance are required. The contact resistance depends on the contact area, the soldering and the pressure applied during soldering. It has been shown that using an optimized contacting method standard deviations of the contact resistance of less than $10\,\text{n}\Omega\,\text{cm}^2$ are achievable [Bay12]. Such contacts require optimized soldering methods and individually contacting of each tape.

5.4.2 Soldering methods

High quality soldering is essential for reproducible contacts with a low contact resistivity. For soldering, the contact areas of copper current leads and the *REBCO* tapes have to be coated with a thin film of solder. The coverage must be absolute, voids have to be prevented as they would reduce the effective contact area. As *REBCO* is degraded by acids, different soldering methods have to be used for the copper current lead and the *REBCO* tapes.

---

[a]Extreme example as multi-stacking is present in that cable. In addition, the contact lengths are with 100 mm significantly too short for 45 tapes.

5.4.2.1 Coating copper with solder

Copper has to be cleaned thoroughly prior to soldering. Grease and other residues have to be removed. Cleaning in four steps distilled water, acetone, ethanol and distilled water, has shown to be effective [Bar09]. After cleaning, the copper current lead is heated to about 150 °C using a temperature controlled heating plate. In50Sn50 solder is applied in the contact areas. To achieve complete coverage in these areas Coloflux[b] fluxing agent is used. Coloflux has shown to be an ideal fluxing agent for use in combination with high temperature superconductors [Hel09]. To apply the fluxing agent a lint-free paper is impregnated with Coloflux. The area which is to be covered in solder is repeatedly touched with the impregnated paper. Solder is applied at the same time and distributed evenly with an soldering iron set to 140 °C. After the contact areas are completely covered in solder, residues of the fluxing agent have to be washed out. To remove the fluxing agent, the copper current lead is to remain on the heating plate. Small amounts of distilled water are applied to the solder and are sponged off with lint-free paper. This has to be repeated several times until all traces of the fluxing agent are removed.

5.4.2.2 Coating *REBCO* tapes with solder

The *REBCO* tapes have to be handled with great care during soldering. To avoid loss of oxygen in the superconducting layer, the temperature of the *REBCO* tapes must be kept below 175 °C [AMS12] at all times. In meander structured tapes, the *REBCO* layer is exposed on the punched out sides. In these tapes fluxing agents can easily degrade the current carrying capabilities. Therefore, it is best to solder *REBCO* tapes for RACC cables without using fluxing agents.

To homogeneously coat *REBCO* tapes with solder, the tapes are first cleaned with distilled water and ethanol. After cleaning, the *REBCO* tapes are heated on a temperature controlled heating plate above the melting point of the solder. For In50Sn50solder, 140 °C have been shown to be ideal. In50Sn50 solder is melted on the heating plate. A soft lint-free paper is pressed into the liquid solder. Small residues of solder stick to the lint-free paper. These residues are now rubbed over the area of the *REBCO* tapes which is to be coated with solder. Light pressure has to be used. This process has to be repeated several times. Oxidation of the surface of the *REBCO* tapes is polished away. The solder itself acts as an abrasive. After several repetitions, the surface is clean and the tapes become coated with solder. If a small well defined area is to be coated with solder, a mask made of paper can be used as shown in **figure 5.20**. Solder is rubbed onto the *REBCO* only in the parts which are accessible through an open area in the mask.

Using this method of soldering [Bay], well defined areas of coated conductor tapes can be coated with solder without exposing the *REBCO* to high temperatures or resorting to fluxing

---

[b]Coloflux is a special mixture of natural/synthetic colophony which is in solution in isopropanol. Coloflux is made by Spirig.

Figure 5.20: Mask made of paper to coat a well defined area of a *RE*BCO tape with solder.

agents. Complete coverage of solder is achieved as oxidation at the surface of the *RE*BCO is abraded during soldering.

5.4.3 Individually contacted tapes

Contacting each tape individually is a solution for improving the homogeneity of the contact resistivity of RACC cables. It is achieved using a stair shaped copper termination; each superconductor tape of a RACC cable is contacted on a different step. A schematic drawing of such a contact is shown in **figure 5.21**. The tapes are soldered with their superconducting layer on bottom to their corresponding step. All steps are of the same dimensions, thus each tape utilizes the same contact area. Through steel cover plates a constant amount of pressure is applied during soldering.

Using the soldering method described in **subsection 5.4.2.1**, the contact areas of the steps are coated with solder. In50Sn50 solder is used. Additionally, the tapes of a RACC cable have to be cut at the correct positions. These positions can be determined by placing the RACC cable on the stair shaped copper current lead with the superconducting layer facing upwards. Beginning at the first step the uppermost *RE*BCO tape of the RACC cable is carefully cut with scissors at the end of that step. At the second step a different tape is now uppermost one. That tape is cut at the end of the second step. This process is repeated until all tapes are of the correct length. It is helpful to fix the RACC cable to avoid shifting. This step in manufacturing stair shaped contacts of RACC cables is shown in **figure 5.22**.

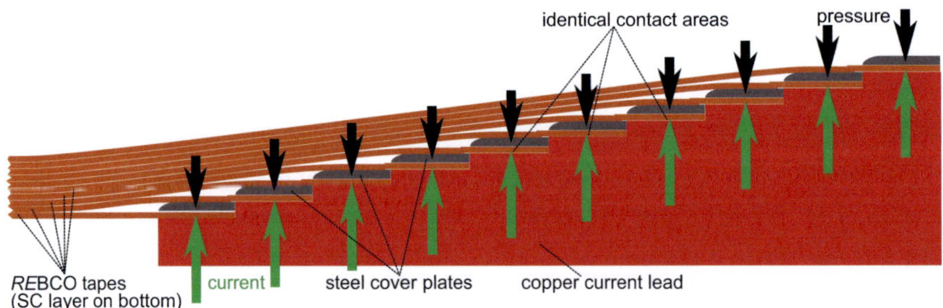

Figure 5.21: Schematic drawing of a stair shaped contact for RACC cables. Each *RE*BCO tape is soldered to a different step. Steel cover plates are used to apply a constant amount of pressure during soldering.

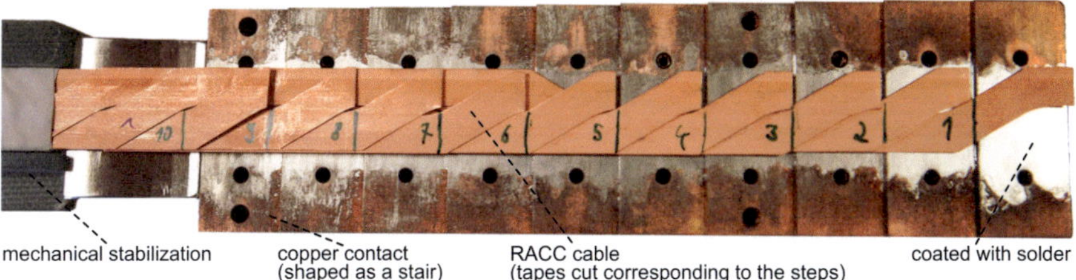

Figure 5.22: First step in assembling a stair shaped contact for RACC cables. The tapes of the RACC cable are cut to their correct lengths. The contact areas on the steps of the copper current leads are coated with solder.

In a further step, the contact areas of the *RE*BCO tapes of the RACC cable have to be homogeneously coated with In50Sn50 solder. To protect the *RE*BCO this should be done with the soldering method described in **subsection 5.4.2.2**. A mask is needed for the solder to be only applied in the contact areas of the tapes. It is essential the soldering is done very carefully to avoid damaging the *RE*BCO. A RACC cable for which this soldering step has been performed is shown in **figure 5.23**.

Figure 5.23: Second step in assembling a stair shaped contact for RACC cables. The contact areas of the tapes of the RACC cable have to be coated with solder.

The stair shaped copper current lead and the tapes of the RACC cable have now been prepared. To solder them together, the stair shaped copper current lead is heated to about 150 °C on a heating plate. Starting with the lowest step additional In50Sn50 solder is deposited on the contact area. The corresponding *RE*BCO tape is placed with the side which has been coated with solder onto the step. A thin steel cover plate is used to fixate the *RE*BCO tape. Using a torque screwdriver a defined pressure is applied. With a lint-free paper excess solder can be removed. This process, which is shown in **figure 5.24**, is repeated until all tapes are soldered to the copper current lead. Controlling the torque is essential as the contact resistance is influenced by the used pressure during soldering. For contacts in the $100\,\mathrm{n\Omega}$ range pressures during soldering of at least $3\,\mathrm{N\,m^{-2}}$ are required [Str12].

As shown in **figure 5.25**, a cover can be attached to protect the contacts. With such a cover, pressure contacts are possible. This allows easy, low resistance connection to the current leads of applications or test facilities.

Figure 5.24: On each step of the stair shaped copper current lead the associated *RE*BCO tape is soldered. With steel cover plated the tapes are fixated. Using a torque screwdriver constant pressure is applied to each contact.

Figure 5.25: A cover can be attached to protect the contacts.

5.4.4 Performance of individually contacted RACC cables

A RACC cable with 10 strands has been constructed. The strands are punched from 12 mm wide *RE*BCO from SuperPower (SP-KIT-20110913-10). All tapes are individually contacted as described in **subsection 5.4.3**. Critical current measurements in a liquid nitrogen bath have been performed. The electric field at each tape and the voltage drop between the copper contacts was registered. As shown in **figure 5.26**, the cable behaved homogeneously. The transition from superconducting state to normal conduction occurred simultaneously for all the tapes. With the $1\,\mu V\,cm^{-1}$ criteria (common for HTS cables [LLG11, TMB11, GFK09]), a critical current of 1.21 kA was achieved at 77 K in self-field conditions. From the ohmic part of the voltage drop over the copper contacts, the contact resistivity of the cable is calculated. An overall contact resistance of 48 nΩ was gained.

To determine the homogeneity of the contacts of the RACC cable, the n-values of each tape and the voltage drop over the contacts are calculated. This is done in a double logarithmic display of the electric field and transport current dependence. In this display the n-value of a tape is the slope of the corresponding curve in the area of E_c to $10 \cdot E_c$ [TH05]. The calculated n-values are in the range of 20.1 to 23.5 for all the tapes. For the whole cable, which is shown as "overall"

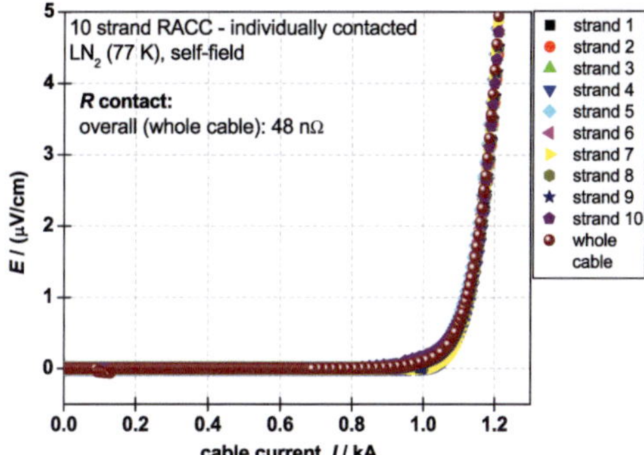

Figure 5.26: Electric field vs. current of a 10 strand RACC cable. Electric field of each tape is shown as "tape 1-10". "Overall" gives the voltage drop over the copper contacts. The ohmic part has been subtracted.

in **figure 5.27**, a n-value of 21.3 is achieved. These n-values corresponds well to data available from the manufacturer of the *RE*BCO tapes (SuperPower). According to SuperPower n-values of their *RE*BCO tapes are in the 21 to 37 range [Sel11b, p. 16]. Individual contacting allows RACC cables retaining the n-value of the single tapes. Even the voltage drop over the contacts, which behave similarly to an averaging of the performance of all the tapes, exhibits a n-value compatible to the original tapes.

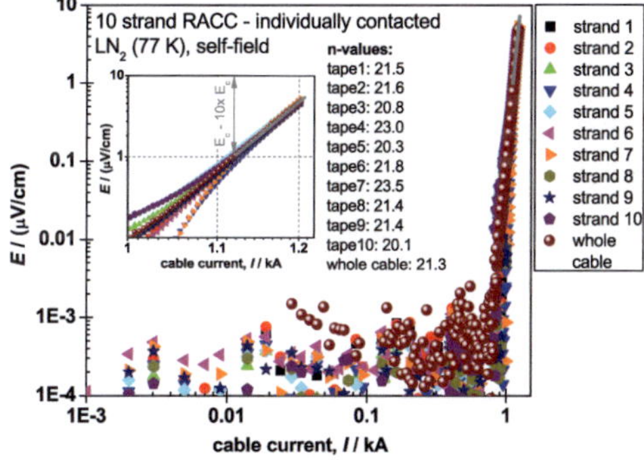

Figure 5.27: Double logarithmic display of the electric field and transport current dependence. In the range E_c to $10 \cdot E_c$ the slope of a curve is the n-value of the corresponding superconductor.

5.4.5 Summary and recommendation

In RACC cables, the *RE*BCO layers are accessible from the outside. Fluxing agents should be avoided as they can enter the *RE*BCO layer and degrade the current carrying capabilities of the tapes. It has been shown that it is possible to completely cover selected areas of *RE*BCO tapes with indium-tin based solder without the use of fluxing agents.

A RACC cable with individually contacted tapes has been manufactured and tested. Test data suggests uniform contact resistance of all the tapes. All the tapes transition from the superconducting state to normal conduction simultaneously, at the same current. An overall contact resistance of $48\,\text{n}\Omega$ was gained. N-values for each tape and for the whole cable have been calculated. The deviations between the n-values are low. All n-values are within the range of the data specified by the manufacturer for the material used.

Individual contacting has shown to be a promising alternative to conventional RACC contacts.

6 High temperature superconductor fusion magnet cable concepts

At present, there are five major HTS cable concepts for fusion magnets which are being developed at companies and research institutions all over the world [MBT11]. The cable concepts are:

Roebel Assembled Coated Conductor (RACC) cables

RACC cables are being developed at the Karlsruhe Institute of Technology (KIT), Germany, and Industrial Research Limited (IRL), New Zealand [GNK06, Lon11]. This HTS cable type is already introduced in **chapter 5**, in which its geometry and contacts are optimized. In **section 6.2** the performance in magnetic background fields of mechanically unstabilized and stabilized RACC cables is investigated.

Coated Conductor Rutherford Cables (CCRC)

CCRCs are being developed at the Karlsruhe Institute of Technology (KIT), Germany [SDR10]. This cable type is described in **section 6.3**. At present there are no CCRC samples available suitable for in-field measurements. Instead, 77 K, self field performance is used as base.

Conductor on Round Core (CORC) cables

CORC are being developed at Advanced Conductor Technologies, USA, and the University of Colorado, USA [Laa09]. CORC cables are investigated in **section 6.4**.

Twisted Stacked-Tape Cables (TSTC)

TSTCs are being developed at the Massachusetts Institute of Technology, USA [TCBM11, TMB]. **Section 6.5** is devoted to this cable concept.

Round Strands Composed of Coated Conductor Tapes (RSCCCT) cables

This cable concept is proposed by the Centre de Recherches en Physique des Plasmas (CRPP) of the École Polytechnique Fédérale de Lausanne (EPFL). This cable concept is described in **section 6.6**. No field dependent measurements are available, 77 K, self-field data is used as base instead.

All cable concepts are described in detail. The mechanical properties, performance at different temperatures and magnetic background fields and the scaling to currents relevant for fusion magnets are investigated as far as possible:

- The mechanical properties of the cables are discussed in **subsection 6.2.2**, **subsection 6.4.2** and **subsection 6.5.2**. Different load directions are considered and assessed.

- The in-field performance is analyzed with magnetic field and temperature dependent measurements in **subsection 6.2.3**, **subsection 6.4.3** and **subsection 6.5.3**. Similar boundary conditions are used for all concepts. The characterization is performed in magnetic background fields up to 12 T and cable temperatures in range of 4.2 K to 80 K, covering possible operation scenarios of future fusion magnets. The setup used for these experiments is described in **section 6.1**.

- The scaling to currents relevant for fusion magnets is investigated in **subsection 6.2.4**, **subsection 6.3.4**, **subsection 6.4.6**, **subsection 6.5.5** and **subsection 6.6.2**. Scaling laws are compiled. The maximal current carrying capabilities of each concept are assessed within the twist pitch limit of the ITER toroidal field coil conductors for different temperatures of the HTS cables.

- Homogeneous contact resistances of all tapes are essential for superconducting cables as shown for example in **section 5.4** by comparing two contacting methods for RACC cables. For all HTS cable concepts, possible contacting methods are described and evaluated in **subsection 6.2.5**, **subsection 6.3.5**, **subsection 6.4.7**, **subsection 6.5.6** and **subsection 6.6.3**.

The experimental setup and the investigations on RACC cables, CCRCs and CORC cables are also published by the author in [BDF12b] and [BDF12a].

6.1 Experimental setup

The FBI (force - field - current) superconductor test facility of the Institute for Technical Physics (ITEP) of the Karlsruhe Institute of Technology (KIT) is used to characterize HTS cables. This test facility (see **subsection 6.1.1**) was previously used for the low temperature superconductor cables [Wei07]. It has been enhanced with a temperature variable insert [BWGS11] to allow testing at sample temperatures above 4.2 K (see **subsection 6.1.2**).

6.1.1 The FBI test facility

The name "FBI" of the test facility is an abbreviation of the possible test parameters, **F** stands for force, **B** for magnetic field and **I** for current. This test facility consists of:

- a superconducting split coil magnet, delivering magnetic background fields up to 12 T. The field is oriented perpendicular to the sample axis and the current. The homogeneous field region (97 % of peak field) is 70 mm long. The sample fits in the gap of the magnet. Above and below the magnet, there are force grips and current leads connecting the sample to the tensile machine and the power supply.

- a tensile machine, applying tensile loads up to 100 kN.

- a low noise DC power supply which can provide up to 10 kA.

- sample, magnet and current leads are cooled with liquid helium in a bath cryostat.

The FBI setup is shown schematically in **figure 6.1**.

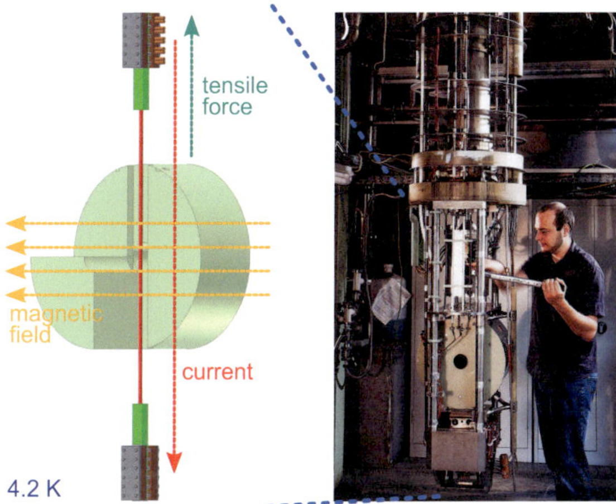

Figure 6.1: Schematic drawing of the superconductor test facility "FBI". It allows critical current measurements on superconductor cables depending on magnetic background fields and tensile loads. Picture from M. Breig.

Samples of 1.1 - 1.2 m length can be tested in this test facility. With the FBI test arrangement, mechanical stresses and strains are applied directly to the sample in contrast to "Pacman"[GDM04] or "Walter Spring"[WDT86] systems. Simultaneous mechanical and electromagnetic investigations are possible [Wei07]. Due to the superconducting magnet (Nb_3Sn) and the superconducting current leads (NbTi) the operating temperature of the test facility is fixed at 4.2 K. However, for the characterization of high temperature superconductor cables, the sample temperatures above 4.2 K are also of interest.

6.1.2 Temperature variable insert

To expand the measurement capabilities of the FBI test facility, a temperature variable insert is constructed to increase the sample temperature while the rest of the test facility remains at 4.2 K.

6.1.2.1 Layout

To raise the sample temperature it is heated while immersed in the liquid helium bath. Thermal insulation between the sample and the bath is necessary to reach temperatures significantly above 4.2 K and to minimize the helium boil off. The temperature variable insert is shown schematically in **figure 6.2**; it consists of:

- Cernox™ temperature sensors [Laka] in direct contact with the sample to determine its surface temperature. Cernox sensors are used as this sensor type combines a high sensitivity at cryogenic temperatures with a very low magnetic field dependence.

- heating foils wrapped around the sample. The heating power of the foils is controlled individually. It can be either kept constant or adjusted using PID controllers to achieve the target sample temperatures.

- thermal shielding with a chamber containing gaseous helium during operation of the temperature variable insert. This shielding is achieved with a two wall construction out of glass-fiber-reinforce-plastics G10. The chamber fits tightly around the sample and is sealed at the upper and lower end to minimizes the exchange of helium.

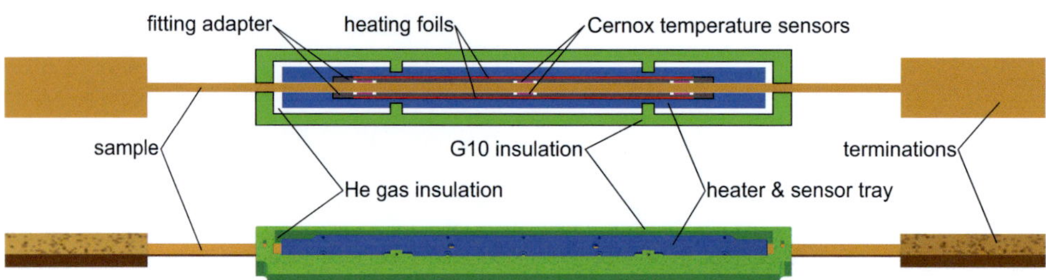

Figure 6.2: Temperature variable insert for the FBI test facility. The insert allows testing at sample temperatures from 4.2 - 77 K. Schematic drawing in side view (upper), technical drawing (lower).

The temperature variable insert fits inside the gap of the magnet of the FBI test facility. It can be used either with round samples up to 10 mm diameter or flat samples of no more than 12 mm width and 5 mm thickness. For each sample geometry, fitting adapters are necessary to ensure good thermal insulation. With this system, the sample temperature can be adjusted from 4.2 K to 77 K. To get current into the sample, both ends of the sample are connected to the current leads of the test facility which remain immersed in the liquid helium bath. Therefore, there can be significant temperature differences in the sample between its contacts and the heated section. This means that during operation of the temperature variable insert, heat is always conducted along the sample from the inside of the temperature variable insert to the helium bath.

The space in the gap of the magnet is used completely by the temperature variable insert. Therefore it is not possible to simultaneously test temperature dependence and mechanical strain dependence. Either the tensile machine, or the temperature variable insert can be used during one measurement. In **figure 6.3**, these extended measurement capabilities are shown schematically.

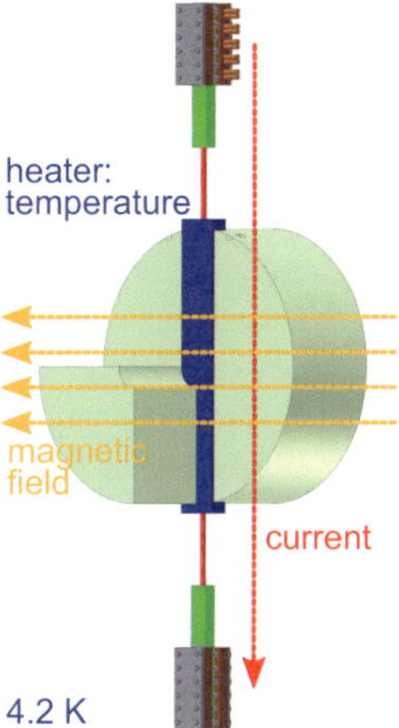

Figure 6.3: Schematic drawing of the superconductor test facility "FBI" with the temperature variable insert. The system allows critical current measurements on superconductor cables depending on magnetic background fields and temperatures.

6.1.2.2 Validation

To verify the performance (stability, response time and temperature distribution) of the temperature variable insert, a superconductor cable dummy is constructed. The dummy consists of a copper rod of 8 mm diameter inside a round stainless steel jacket with 1 mm of wall thickness. This composition simulates cable-in-conduit-conductors, with a high thermal conductivity in the cable area (the copper rod) and a lower thermal conductivity in the jacket (the stainless steel tube). Four Cernox sensors are used to measure the temperature of the dummy at different positions. The layout of the dummy and the placement of the sensors is shown schematically on the left of **figure 6.4**. Two temperature sensors are embedded in the center of the copper rod (sensor positions 2 and 4). Sensor 2 is located in the center of the heating section while sensor 4 is at the upper border of the heating section. The other two sensors are attached to the stainless steel jacket to measure the surface temperature of the dummy. Again, one sensor (sensor position 1) is in the center of the heating section while the other one is located at the upper border (sensor position 3). On the right of **figure 6.4**, the readout of these sensors is shown for different heating powers of the temperature variable insert.

These temperature sensors of the cable dummy are used to assess the performance of the temperature variable insert. Response time, stability and temperature distribution are analyzed.

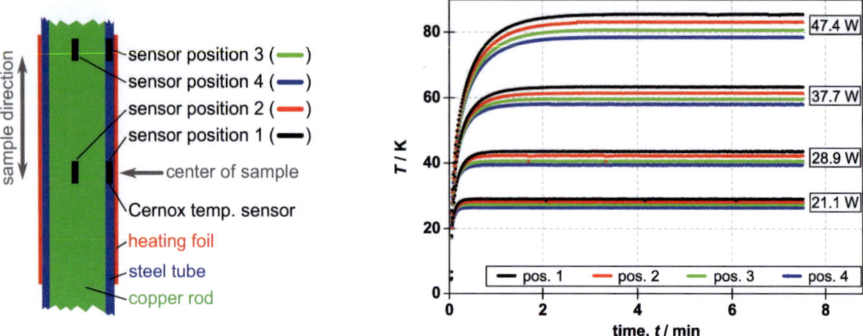

Figure 6.4: Validation of the temperature variable insert with a CICC dummy. The dummy consists of a copper rod inside a steel tube. Four Cernox temperature sensors are placed at different positions. Temperature sensor positions and the layout of the dummy are shown on the left. The time dependent readout of the sensors is show for different heating powers on the right.

Response time

The response of the temperature variable insert is fast, especially in the temperature range from 4.2 K to 30 K, due to the low heat capacity of the materials at these temperatures. In this temperature range it takes less than 1 min to heat up the sample and reach a stable temperature. At higher temperatures, the thermal capacities and the heat loss through thermal conduction along the sample are increased. Therefore, far more thermal energy has to be deposited in the sample to reach thermal equilibrium. Near 80 K, the heating up time to reach a stable temperature is increased to 3 min.

Stability

After the heat up time, the measured temperatures are stable with flat temperature vs. time curves. Temperature variations at constant heating power remain below 0.5 K at all investigated temperatures and can be neglected. Thus, sample temperature are assumed to be constant after the heat up time.

Temperature distribution

The cable dummy is surrounded with heating foils inside the temperature variable insert. Outside the insert, the dummy remains immersed in the liquid helium bath. If the heating foils are turned on, there are significant temperature differences in the sample. Heat is transported along the sample from the heated section to the helium bath. This heat flux depends on the temperature difference, the distance, the cross sectional area of the sample and the thermal conductivities of the constituent materials. Consequently the sample temperature is not homogenous; there is always a temperature gradient in the sample. In the center of the heated section (sensor positions 1 & 2) the temperatures will be highest. Near the border of the temperature variable insert (sensor positions 3 & 4) the temperatures are lowest.

The cable dummy is heated on its surface, through the stainless steel jacket. The thermal conductivity of the jacket material is much lower compared with the conductivity of the central copper rod. Thus, the thermal conductivity of the cable is significantly higher in longitudinal direction than in radial direction. Along the sample (longitudinal direction) the main part of the heat flux takes place in the copper center, resulting in a radial temperature gradient. The temperature in the heated surface (sensor positions 1 & 3) is higher than in the center (sensor positions 2 & 4) of the cable dummy.

The precedence of the temperatures remains true for all investigated heating powers. At increased temperatures, the differences become more pronounced. At around 28 K the temperature differences are less than ± 2 K; near 80 K, the temperature differences are of the order of ± 5 K.

6.1.3 Assessment

Tests with the CICC cable dummy verified the principle operation of the temperature variable insert. With the insert it is possible to reach sample temperatures up to 80 K and operating the insert does not disrupt the magnet. It is still possible to reach background field of 12 T even at maximal heating power. No instabilities are observed if the heating power is kept below 50 W. The temperature variable insert works fast and stable, especially at temperatures in the 4.2 - 30 K range. However, due to the thermal conduction along the sample, the temperature of the sample inside the insert is not homogeneous. There is a distinct temperature variation within sample especially at high heating powers.

For measurements with HTS cables, the inhomogeneity of the cable temperature has to be considered. In an HTS cable it is not possible to place temperature sensors inside the conductor due to the very limited space. This precludes an experimental approach. Instead, the temperature distribution is calculated using Finite-Element-Method (FEM) models with the commercial software package "Comsol"[a]. As the temperature distribution is dependent on the geometry and the material properties of the sample, this calculation has to be done for each HTS cable type individually. These calculations are presented in **subsection 6.4.5** and **subsection 6.5.4**.

The temperature variable insert extends the measurement capabilities of the FBI test facility; the enhanced facility is ideal for the characterization of high temperature superconductor cables.

6.2 Roebel Assembled Coated Conductor (RACC) cable

This HTS cable concept was first demonstrated in 2006 with an HTS cable consisting of 16 commercial DyBCO strands carrying 500 A at 77 K in self-field [GNK06]. To date, the RACC

---

[a]Comsol Multiphysics from http://www.comsol.com/

cable concept has been improved to 45 strands carrying 2.6 kA (77 K in self-field). RACC cables are mainly developed by the Karlsruhe Institute of Technology (KIT) and Industrial Research Limited (IRL).

6.2.1 Composition

Roebel assembled coated conductor cables have already been introduced in **chapter 5**. This HTS cable type consists of meander shaped coated conductor tapes, which are assembled using the Roebel method, resulting in cables of rectangular geometry with the same width as their constituent *RE*BCO tapes. The thickness of the cable and the twist pitch depends on the number of tapes. In **figure 5.3**, a RACC cable of five tapes is shown schematically.

6.2.2 Mechanical properties

In fusion magnets, due to currents in tens of kilo amperes range and backgrounds fields of more than 10 T, the Lorentz forces on the conductor are significant. Thus, the mechanical properties of conductors for fusion magnets are critical. In RACC cables, the mechanical properties are anisotropic. There are three different load directions:

Mechanical loads parallel to the cable direction
This is the load direction of the hoop stresses in magnet conductors. In RACC cables hoop stress leads to longitudinal tensile stress on the *RE*BCO tapes. With critical stress limits of 500 - 700 MPa, the bare tapes are insensitive to stresses in this direction. However, the meander structure necessary for the Roebel assembly results in areas of the tapes in which stresses are concentrated. These areas are more sensitive to mechanical loads, limiting the overall mechanical properties of RACC cables. However, a 15x5 (15 strands with 5 mm width each) RACC still retained 260 MPa critical longitudinal tensile stress [b].

Mechanical loads perpendicular to the cable and perpendicular to the surface of the tapes
The forces press on the flat side of the conductor and compress the tape. In this orientation *RE*BCO tapes from SuperPower are able to withstand more than 240 MPa pressure (see **subsection 2.3.3.2**).

Mechanical loads perpendicular to the cable and parallel to the tape surface
In this orientation the forces press on the thin side of the tapes. It can be expected that *RE*BCO tapes are sensitive to mechanical loads in this direction and may be damaged easily. Thus, this load orientation is the "worst case" force scenario for RACC cables. Therefore, the mechanical properties of RACC cables must be examined closely in this direction.

---

[b] calculated with data presented in [BBL12]: 1.95 kN critical tensile load of the whole cable $\Rightarrow$ 130 N per tape (5×10^{-7} m^2) $\Rightarrow$ critical longitudinal tensile stress of 260 MPa.

As the Lorentz forces ($\vec{F}_L \sim \vec{I} \times \vec{B}$) are perpendicular to both the current direction $\vec{I}$ and the direction of the magnetic field $\vec{B}$, the worst case load orientation arises with currents parallel to the cable and magnetic fields perpendicular to the tape surface as shown in **figure 6.5**. Magnetic fields perpendicular to the tape surface ($B \perp$ tape surface) are most demanding on RACC cables and are their "worst case" field orientation. This orientation is used for the in-field characterization of RACC cables as the mechanical properties of the cable can be expected to be most critical.

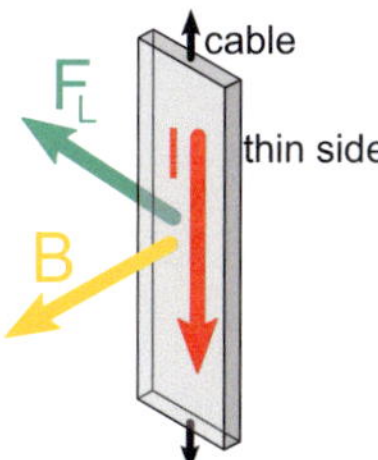

Figure 6.5: Worst case field orientation of RACC cables. Magnetic field is perpendicular to the tape surface resulting in Lorentz forces on the thin side of the *REB*CO tapes.

6.2.3 Field and temperature dependent measurements

Two RACC cable samples are prepared for temperature and field dependent measurements:

- a 10 strand RACC cable manufactured by KIT. It is inserted in an adapter of the temperature variable insert fitting for rectangular samples. The adapter consists of two pieces, a U-shaped block of glass-fiber-reinforced-plastic (G10) and a T-shaped cap of the same composite material. These fit around the cable preventing movement and providing a basic level of support in the magnetic field. As this sample is neither clamped tightly nor impregnated with glue, resin or solder, the strands of the cable are not fixated completely. Lorentz forces act on each strand individually and can cause movement of the tapes.

- a 15x5 (15 strands with 5 mm width each) RACC cable manufactured by IRL. To provide high mechanical stabilization against Lorentz forces, this cable is completely impregnated with an Araldite epoxy resin and quartz powder mixture as described in **subsection 4.2.3**. The epoxy resin glues the cable into the G10 support, prevents movement and connects the strands. Movement of individual tapes is suppressed. This cable can be expected to exhibit superior mechanical properties.

Both samples exhibit comparable performance in liquid nitrogen (77 K, self-field) prior to in-field measurements as shown in **figure 6.6**. The critical current of the 10 strand RACC without epoxy impregnation is 1.12 kA at 1 μV cm^{-1} and 1.20 kA at 5 μV cm^{-1} with n-values of 21.3

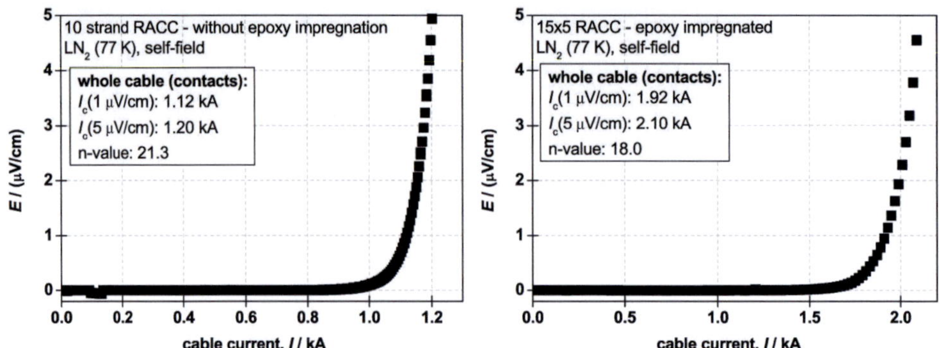

Figure 6.6: Electric field vs. current of RACC cables: an individually contacted 10 strand RACC without epoxy impregnation (left) and an epoxy impregnated 15x5 RACC (right).

corresponding to an effective critical current density of 200 A/cm−width. The epoxy impregnated 15x5 RACC is able to carry $1.92\,\text{kA}$ at $1\,\mu\text{V cm}^{-1}$ and $2.10\,\text{kA}$ at $5\,\mu\text{V cm}^{-1}$ with n-values of 18.0. Its effective critical current density is with 250 A/cm−width higher.

The cables are characterized in worse case field orientation ($B \perp$ tape surface) in magnetic background fields from zero field to 12 T at different temperatures using the temperature variable insert of the FBI test facility.

6.2.3.1 10 strand RACC - no epoxy impregnation

As shown in **figure 6.7**, the 10 strand RACC cable (no epoxy impregnation) exhibits distinct superconducting transitions at self-field and low magnetic background fields (2 T and 4 T). The electric field vs. current curve begins to deviate from their ideal shape at 6 T magnetic background field as some degradation is already occurring in the cable. At 8 T and above the

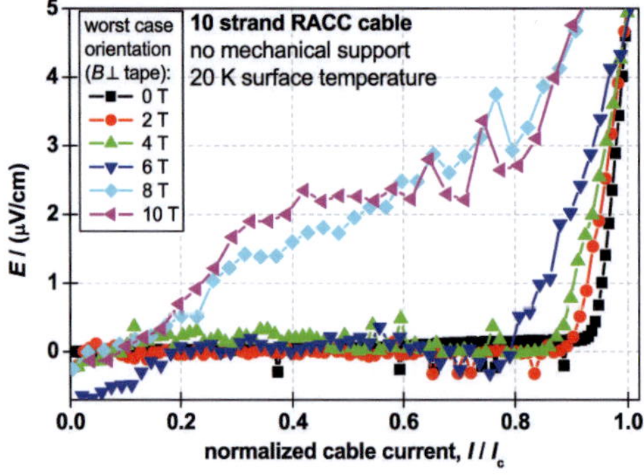

Figure 6.7: Electric field vs. normalized current curves of a 10 strand RACC cable at different magnetic background fields. Cable measured in worst case field orientation ($B \perp$ tape surface) without additional mechanical support (no epoxy impregnation).

Figure 6.8: Damaged sections of the 10 strand RACC (no epoxy impregnation) after in-field
measurements in worst case field orientation.

sample is almost completely degraded. It is not possible to determine a critical current, there
is no distinct transition visible between the superconducting and the normal conduction states.
Even if the magnetic background field is reduced again, the degradation remains. The cable has
been irreversible damaged in the tests in background field above 6 T due to the Lorentz forces
which press on the thin sides of the tapes. The damage of the cable is clearly visible as shown in
figure 6.8. There are tears in the inner radii of the crossing sections of the *RE*BCO tapes. The
tapes are severely damaged, the cable does not exhibit any superconducting properties, even at
4.2 K, self-field.

It is clear that RACC cables are very sensitive the mechanical loads in this orientation. Additional mechanical support is necessary to protect the cable in high magnetic background fields.
Impregnation with epoxy resin is investigated as an option to provide this support.

6.2.3.2 15x5 RACC - epoxy resin impregnation

As described in **subsection 4.2.3**, a 15x5 RACC from IRL is impregnated with a mixture of
Araldite epoxy resin and quartz powder (mixing ratio 1:1) to provide the cable with additional
mechanical support. At 77 K, self-field no degradation of the current carrying capabilities is
observed prior and after impregnation.

The magnetic field and temperature dependence of the critical current of this sample is
measured in the FBI test facility. In contrast to other HTS samples, a new split-coil magnet is
used in this test. The high field region (97 % of peak field) of the new magnet is 92.2 mm long.
The temperature variable insert is adjusted to fit into the new magnet. To verify the mechanically
stabilizing properties of the epoxy resin impregnation, it is planned to increase to magnetic
field up to 12 T while measuring the critical current of the sample after each 2 T increase. After
reaching maximal field, the field will be slowly reduced to zero while measuring the critical
currents. The current carrying capabilities during field increase and decrease will be compared.
This procedure will be repeated several times to check for degradation of the current carrying
capabilities due to the Lorentz forces at the high magnetic fields.

However, the sample was already damaged at zero magnetic background field during a test
run of the temperature variable insert. With the insert, the surface temperature of the cable

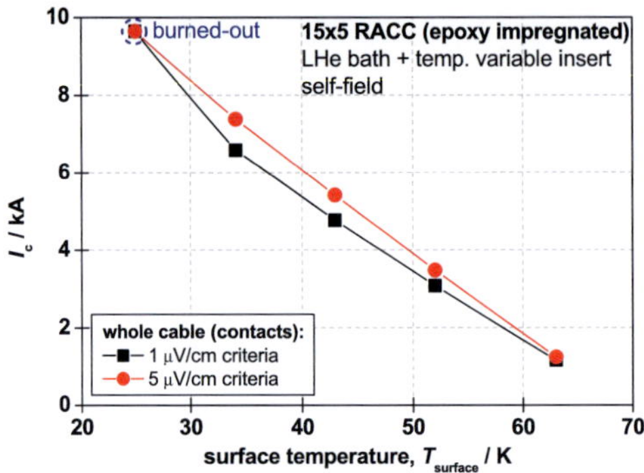

Figure 6.9: 15x5 RACC from IRL tested with the temperature variable insert in the FBI test facility. The sample surface temperature was reduced from 63 K to 4.2 K. At 25 K surface temperature the sample burned out during a quench at a current of 9.65 kA.

is reduced in several steps from 63 K to 4.2 K. As shown in **figure 6.9**, the current carrying capabilities are determined at each step using the $1\,\mu\mathrm{V\,cm^{-1}}$ criteria and the $5\,\mu\mathrm{V\,cm^{-1}}$ criteria. At a surface temperature of 25 K, the cable burns out at 9.65 kA, preventing the previously explained magnetic field cycles (Lorentz force degradation measurements).

Cause of the burning out of the 15x5 RACC sample is a combination of three main factors:

Thermal insulation

The 15x5 RACC sample is designed for optimal mechanical stabilization. It is inserted into a G10 former and completely impregnated with a epoxy resin and quartz powder mixture. This mixture and the glass-fiber-reinforced plastic G10 are thermal insulators with thermal conductivities of $< 0.2\,\mathrm{W\,K^{-1}\,m^{-1}}$ in the relevant temperature range (see **figure 4.17**). Thus, the cable is not in direct contact with the cooling bath. Heat has to be transported through the epoxy resin mixture and the G10 former limiting the effective cooling contact surface. During the transition from superconducting state to normal conduction, heat is generated in the cable. Due to the bad thermal link to the cooling bath, the temperature of the sample increases. Higher sample temperature means lower critical current, increasing the electric fields and the heat generation. In a chain reaction, this speeds up the superconducting transition.

Electrical stabilization

The sample does not include additional electrical stabilization. As the sample was optimized for mechanical properties, the available cross sectional area is completely used for the mechanical stabilization. Only the copper in the SuperPower *REBCO* tapes electrically stabilizes the cables, and has to carry the current during and after the transition to normal conduction. The copper cross sectional area of the 15x5 RACC is $3\,\mathrm{mm^2}$ and is small, resulting in a very high current density of $3.27\,\mathrm{kA\,mm^{-2}}$ in the copper after the quench of the sample.

Response time

The response time of the FBI test facility, from the quench of the sample to the complete shutdown of the current, is approximately 200 ms. Due to the very high current density in the copper ($3.27 \, \mathrm{kA \, mm^{-2}}$), bad thermal link to the cooling bath and low heat capacity of the materials at cryogenic temperature, the temperature of the sample increases drastically in this time period. During a quench of the sample at nearly full current (10 kA), temperatures can be reached which destroy the *RE*BCO tapes and result in the burn out of the sample.

To prevent the burn out of the sample, several steps have to be introduced. It is possible to improve the response time of the test facility by increasing the scanning frequency of the digital quench detection system from currently 50 Hz to 200 Hz. Analog quench detectors, with slightly higher detection thresholds can be used in parallel, increasing the swiftness and the reliability of the detection. At present, the relays combining all sources of faults and interrupts to initiate emergency shutdowns of the power supply and the dump of the current are electromechanical. Replacing these with solid state relays will reduce the switching times significantly.

These changes however can only reduce the response time so far. The power supply itself requires several milli seconds to dump the current and switch from maximal current to zero in an emergency shutdown. Thus, the more important changes are on the sample, especially if such a cable represents a conductor of a large magnet. In a fusion magnet, there are several GJ of energy stored withing the magnetic field, limiting the shutdown speed. Therefore, to allow safe operation of an HTS fusion magnet, significant additional electrical stabilization is indispensable. The current levels and the stored energies are far too high to be carried solely by the copper stabilization of the *RE*BCO tapes. All other HTS cables, tested in liquid helium with the FBI test facility (CORC cable in **section 6.4** and the TSTC in **section 6.5**), therefore contain additional copper for electrical stabilization.

A suitable amount (cross sectional area) of electrical stabilization must be included in any fusion magnet conductor (see **section 6.8**). Additionally, due to low thermal conductivity of glues and resins, the influence of epoxy impregnation on the cooling efficiency has to be considered, too.

6.2.3.3 In-field tests at CERN

RACC cables have been tested from CERN in different magnetic field orientations [FBBT12]. Samples include a 10 strand RACC (consisting of 10 SuperPower tapes each 6 mm wide) from KIT and a 15x5 RACC (15 SuperPower tapes, 5 mm wide) from General Cable Superconductor (GCS). The samples are cooled to 4.2 K in a liquid helium bath. A 10 T NbTi dipole magnet is used to generate magnetic background fields. The current carrying capabilities of these samples are measured at different magnetic fields and orientations. Two field orientations are used, magnetic fields parallel to the surface of the RACC cables (referred to as parallel fields) and

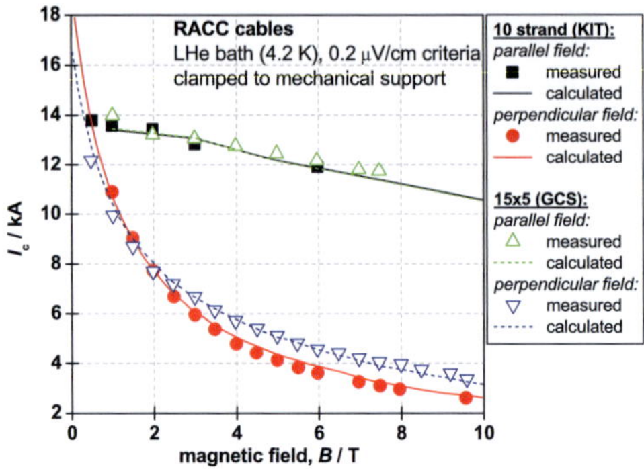

Figure 6.10: Critical currents of RACC cables in magnetic background fields of different orientations at 4.2 K. Samples are from KIT (10 strand RACC) and General Cable Superconductor (GCS). Measured values are shown as points while calculated data shown as lines. Measurements performed at CERN in the FRESCA test facility. Data from [FBBT12].

magnetic fields perpendicular to the RACC cable surface (referred to as perpendicular fields). In parallel field orientation, the Lorentz forces, which are always perpendicular to magnetic field and current, press on the thin sides of the samples. This is the previously introduced "worst case" field orientation (see **figure 6.5**).

At CERN, the RACC cables are clamped between two rigid stainless steel plates for mechanical stabilization, applying transversal pressures of approximately 40 MPa to the samples. The clamping fixes the cables and prevents movement of individual *REBCO* strands during the in-field tests. In **figure 6.10**, the measured data (shown as points) is compared with extrapolated data (shown as lines).

The samples behave similarly to the extrapolations. A $0.2\,\mu\mathrm{V\,cm^{-1}}$ criteria is used to determine critical currents. No degradation is observed, even in "worst case" field orientation. The tested RACC cables are to carry up to 11.7 kA in parallel magnetic fields of 7.5 T. In perpendicular field orientation, critical currents up to 3.6 kA were achieved at 9.6 T [FBBT12].

These tests verify the applicability of the RACC cable concept at high magnetic background fields. Sufficient mechanical stabilization is necessary to prevent the cables from being damaged by the Lorentz forces especially in the "worst case" field orientation. Clamping RACC cables into a support structure of rigid structural material has been shown to be a valid method to provide this mechanical support.

6.2.4 Scaling to currents relevant for fusion magnets

For fusion magnets, the conductor has to carry tens of kilo amperes at magnetic background fields of 12 T or more. Thus, the current carrying capabilities of RACC cables have to be extended drastically compared with existing samples, for a conductor suitable for fusion magnets. This results in a drastic increase of the number of tapes of at least one order of magnitude compared with the largest existing samples.

However, increasing the number of tapes influences the minimal possible twist pitch. RACC cables are transposed completely. During one twist pitch each tape is in each position. The complete transposition is an intrinsic advantage of RACC cables compared with the other HTS cable concepts. Minimizing the areas which are penetrated by magnetic fluxes and homogenizing the currents carried in each tape. This results in low AC losses, with small coupling currents between the strands [TVP10].

If the ITER magnet system (toroidal field coils) is used as a reference, the conductors of fusion magnets must be able to carry currents of at least 68 kA in magnetic background fields of 12 T. With the present *REBCO* superconductor performance, several hundred tapes are required to reach these currents. In the following, RACC cables are extrapolated to satisfy the magnetic fields and currents of the ITER toroidal field coil conductors.

6.2.4.1 Extrapolation with present *REBCO* tape performance

The extrapolations assume:

- a current in the cable of 68 kA.

- a magnetic background field of 12 T.

- a negligible self-field effect of the cable at 12 T background field.

- cable temperatures of 4.2 K, 22 K and 50 K.

- current carrying capabilities of 2012 SuperPower *REBCO* tapes: 250 A/(cm−width) at 77 K, self-field (see [Haz12, p. 14]).

- magnetic field and temperature dependence of SuperPower *REBCO* tapes (data taken from [Haz12, p. 8]).

- a degradation of the current carrying capabilities due to punching and cabling of 3 % [GFK09].

- a safety factor I_{op}/I_c of 1. However, a safety factor will later be considered in **section 6.8** in the extrapolation of an HTS winding pack for fusion magnets.

The magnetic field and temperature dependence used in these calculations is shown in **figure 6.11**.

The data is for SuperPower tapes in perpendicular field orientation and has been published by the manufacturer [Haz12, p. 8]. In the following, the required amounts of *REBCO* tape width and the corresponding number of tapes are calculated using the boundary conditions.

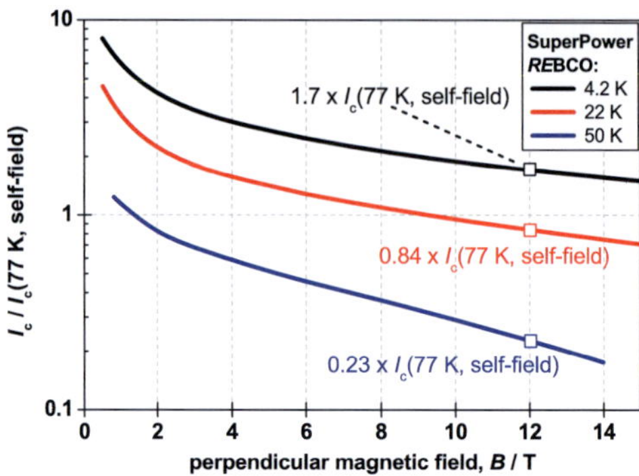

Figure 6.11: Magnetic field and temperature dependence of SuperPower *RE*BCO tapes in perpendicular field orientation. The current carrying capabilities at perpendicular background fields of 12 T are compared with the critical current at 77 K, self-field. Data from [Haz12, p. 8].

6.2.4.2 Corresponding RACC cables

*RE*BCO strands of RACC cables are commonly 2 mm, 5 mm or 6 mm wide. The transposition length per number of tapes depends on the manufacturing process of the cable. In **table 6.1** the transposition length per tape and per superconductor width is given for different RACC cable types manufactured either by KIT or IRL.

RACC cable are scaled to carry 68 kA in 12 T perpendicular background fields at cable temperatures of 4.2 K, 22 K and 50 K, as shown in **table 6.2**.

The minimal transposition lengths τ_{min} are between 3.46 m (KIT - 6 mm wide strands) and 7.43 m (IRL - 2 mm wide strands) at cable temperatures of 4.2 K. For operation at 22 K this is increased to 6.93 m (KIT - 6 mm wide strands) and 14.85 m (IRL - 2 mm wide strands), respectively. Operating temperatures of 50 K would lead to transposition lengths of even 25.95 m (KIT - 6 mm wide strands) and 55.67 m (IRL - 2 mm wide strands). Comparing these lengths to the twist pitch of ITER (toroidal field coil conductor stage 5: 0.42 m $\pm$ 0.02 m [ITE09e, p. 20]), the twist pitch of the RACC cables is at least 8.2 times larger even at 4.2 K cable temperature.

In addition, the height of the RACC cables would be from 13.75 mm (utilizing 6 mm wide

Table 6.1: Transposition length per number of tapes for RACC cables manufactured by KIT (data from [BDF12a]) and IRL (data from [Lon11]).

manufacturer	strand width	transposition length	
IRL	2 mm	9 mm/tape	4.5 mm/(mm−width)
IRL	5 mm	20 mm/tape	4.0 mm/(mm−width)
KIT	2 mm	4.2 mm/tape	2.1 mm/(mm−width)
KIT	6 mm	12.5 mm/tape	2.1 mm/(mm−width)

Table 6.2: Scaling of RACC cables to fusion relevant currents (68 kA at 12 T $\perp$) at cable temperatures of 4.2 K, 22 K and 50 K.

cable temperature T	4.2 K	22 K	50 K
J_c(77 K, self-field)	250 A/(cm–width)	250 A/(cm–width)	250 A/(cm–width)
lift factor: $J_c(T, 12\,\mathrm{T}\,\perp)/J_c$(77 K, self-field)	1.70	0.83	0.23
cabling degradation	3 %	3 %	3 %
$J_e(T, 12\,\mathrm{T}\,\perp)$	412 A/(cm–width)	201 A/(cm–width)	55 A/(cm–width)
effective REBCO width	1.65 m	3.38 m	12.36 m
number of 2 mm strands	825	1690	6180
number of 5 mm strands	330	675	2472
number of 6 mm strands	275	564	2060
cross sectional area $A_{\mathrm{RACC}}{}^{\alpha}$	1.7 cm^2	3.4 cm^2	12.4 cm^2
minimal twist pitch (KIT method) $\tau_{\min}$	3.46 m	6.93 m	25.96 m
fraction of ITER twist pitch (0.42 m $\pm$ 0.02 m)	824 %	1650 %	6181 %

α RACC of 12 mm width (6 mm wide strands).

strands) to 41.25 mm (utilizing 2 mm wide strands) at 4.2 K cable temperature. The height is increased proportional to the number of tapes at 22 K and at 50 K. Because of this, to scale RACC cables to the currents and magnetic fields of fusion magnets, an increase of the REBCO tape performance is necessary. With the current carrying capabilities of today (at 12 T $\perp$), this is not feasible even at 4.2 K cable temperature.

However, more than one RACC cable can be used to carry fusion relevant currents. RACC cables can be used as subcables of a more complex cable design. Such a concept is the Coated Conductor Rutherford Cable (CCRC) which is presented in **section 6.3**.

6.2.5 Contacts

Contacting methods of RACC cables are described in detail in **section 5.4**. The individual contacting of all tapes, with identical contact areas and pressures during soldering, is shown to provide constant contact resistances between the copper and all tapes. Such homogenous contact resistance is essential for good cable performance as it equalizes the current transported in each tape. The proposed "stair" or "double stair" copper contacts are shown to work well with small RACC cables. The scaling of these geometries to several hundreds of REBCO tapes remains a challenge.

6.3 Coated Conductor Rutherford Cable (CCRC)

To reach fusion relevant currents, the RACC design has to be extended allowing much higher number of tapes without increasing the twist pitch. This can be achieved by combining several RACC cables into a larger cable as proposed in [SGG10]. In such a concept the RACC cables are the strands or subcables. These have to be transposed completely as well.

6.3.1 Composition

Coated Conductor Rutherford Cables (CCRCs) are such a concept. Several RACC cables are wound around a former using the Rutherford winding method. Rutherford winding means that the RACC cables are bent in- and out-of-plane around a former of structural material. Bending of RACC cables has been investigated by [SGG10, Kar12] and is shown in **subsection 6.3.2**. Grooves in the former fixate the RACC subcables. The former itself provides mechanical stabilization of the whole cable and cooling channels. For further mechanical stabilization CCRCs can be equipped with a jacket of structural material. This cable concept is shown schematically in **figure 6.12**.

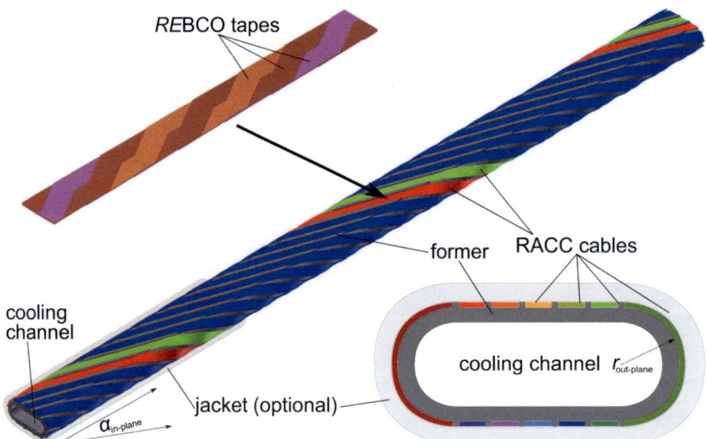

Figure 6.12: Schematic drawing of a Coated Conductor Rutherford Cable (CCRC). Roebel Assembled Coated Conductor (RACC) subcables are wound around a former of structural material in the Rutherford winding method. A jacket of structural material provides additional mechanical stabilization. Picture after [SGG10].

The Rutherford cabling method fully transposes its strands. As the superconductor tapes in the strands, the RACC's subcables, are also transposed entirely, complete transposition of all *RE*BCO tapes is ensured in the CCRC concept.

6.3.2 In- and out-of-plane bending

For Rutherford winding, RACC cables are bent in- and out-of-plane simultaneously as shown schematically in **figure 6.12**. In-plane, winding angles $\alpha_{\text{in-plane}}$ up to $20°$ occur in the proposed CCRC geometry [BDF12a, p. 14]. Out-of-plane, the bending radii $r_{\text{out-plane}}$ depend on the diameter of the former. The impact of in- and out-of-plane bending on the current carrying capabilities has been investigated by [SGG10, Kar12] for 4 mm wide RACC cables consisting of 10 SuperPower strands. Using a specialized bending apparatus [SGG10], critical currents are measured during bends in both directions. Different in-plane winding angles $\alpha_{\text{in-plane}}$, out-of-plane bending radii $r_{\text{out-plane}}$ and orientations of the superconducting layers of the RACC cables are analyzed systematically as shown in **figure 6.13**.

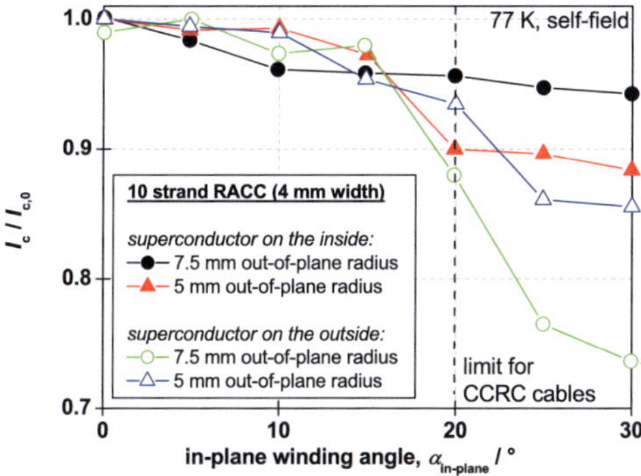

Figure 6.13: In- and out-of-plane bending of RACC cables for CCRC winding. Normalized critical current at different in-plane winding angles $\alpha_{\text{in-plane}}$ (x-axis) and out-of-plane bending radii $r_{\text{out-plane}}$ (legend) with the superconducting REBCO layers in compression (on the inside) or in tensile orientation (on the outside). Data from [Kar12, BDF12a].

Increasing the in-plane winding angle $\alpha_{\text{in-plane}}$ increases the degradation of the current carrying capabilities. At $20°$, which is the limit for the proposed CCRC geometry [BDF12a, p. 14], degradations from 5 % (7.5 mm out-of-plane bending radius, superconductor to the inside) to 12 % (5.0 mm out-of-plane bending radius, superconductor to the outside) occur. On average, lower out-of-plane bending radii $r_{\text{out-plane}}$ also have a negative effect on the current carrying capabilities. The orientation of the superconducting layers plays a major role. RACC cables with their superconducting layers on the inside exhibit significantly lower degradation of the critical currents for all in-plane winding angles $\alpha_{\text{in-plane}}$ and out-of-plane bending radii $r_{\text{out-plane}}$ as the superconducting REBCO layers are in compression. Similar to transverse tensile stresses (see right of **figure 2.14**), tensile bending stresses (REBCO layers on the outside) exhibit stronger impacts on the current carrying capabilities of SuperPower REBCO tapes.

6.3.3 Current carrying capabilities of a sub-size CCRC

The proposed CCRC geometry has been experimentally verified in sub-size by [Kar12]. Three 4 mm wide RACC cables (10 SuperPower *RE*BCO tapes each) are wound around a CCRC former of 15 mm diameter (corresponds to an out-of-plane bending radius of 7.5 mm). An in-plane winding angle of 20° is used. All tapes of the RACC cables are oriented with their superconducting *RE*BCO layers facing to the inside (facing the former). This setup is shown in **figure 6.14**.

Figure 6.14: Sub-size CCRC consisting of 3 RACC subcables with a width of 4 mm. Picture from [Kar12].

The current carrying capabilities of the straight RACC cables are measured and compared with their critical currents after winding. Following the in- and out-of-plane bending experiments, a degradation of the current carrying capabilities of the RACC cables of about 5 % is to be expected in this configuration. The data is compared in **table 6.3** (data from [Kar12][BDF12a, p. 15]).

Table 6.3: Current carrying capabilities of straight and wound, 4 mm wide, RACC cables used in a sub-size CCRC sample. Data from [Kar12][BDF12a, p. 15].

	straight cable	CCRC wound	degradation	assessment
RACC 1	468.8 A	152.0 A	67.6 %	solder in bent zone ⇒ damaged
RACC 2	464.2 A	423.2 A	8.6 %	degradation slightly larger than expected
RACC 3	463.2 A	58.0 A	87.4 %	solder in bent zone ⇒ damaged

In RACC cable 1 and RACC cable 3 strong degradations occurred near their soldered ends. Solder has penetrated into the bent zone increasing the stiffness of the RACC cables, resulting in damage to RACC 1 and RACC 3 on bending. In RACC 2, the solder was applied correctly. In this cable, the current carrying capabilities are reduced from 464.2 A (straight cable) to 423.2 A. This corresponds to a degradation of 8.6 %, which is slightly larger than the expected 5 % degradation from the in- and out-of-plane bending experiments. The retention of the current carrying capabilities of RACC 2 shows that CCRCs can be realized if the soldering of the RACC subcables is done correctly.

Field and temperature dependent measurements of CCRCs with the FBI test facility are scheduled for 2014 within the European Fusion Development Agreement (EFDA) task.

6.3.4 Scaling to currents relevant for fusion magnets

CCRCs can be scaled to very high currents by increasing the number of RACC subcables and the number of *RE*BCO tapes per subcable. The currents and magnetic background fields of fusion magnets are possible at twist pitches τ shorter than the stage 5 twist pitches of the ITER toroidal field coil conductors ($0.42\,\mathrm{m} \pm 0.02\,\mathrm{m}$ [ITE09e, p. 20]) which are used as a reference for HTS fusion magnets. The twist pitches of all HTS cable concepts are compared with this reference.

The proposed CCRC geometry utilizes 4 mm wide RACC subcables (with 2 mm strand width). Using the KIT RACC cabling method, up to 200 tapes are possible, corresponding to 200 mm effective superconductor width, at a twist pitch comparable to the ITER reference (0.42 m). The effective *RE*BCO width necessary for fusion relevant currents (68 kA at 12 T $\perp$) is calculated using the method and boundary conditions described in **subsection 6.2.4.1**. However, the degradation due to the cabling is increased from 3 % (for the RACC cabling) to 13.3 %[c] to account for the degradations occurring in the RACC and the CCRC cabling stages in the proposed CCRC geometry. The degradation is influenced by the out-of-plane bending radius r_{out} and the in-plane winding angle $\alpha_{\mathrm{in\text{-}plane}}$ as shown in **subsection 6.3.2**. The required amounts of *RE*BCO tape (effective width) and the corresponding number of RACC subcables in the CCRC are calculated for cable temperatures of 4.2 K, 22 K and 50 K. The minimal twist pitch τ_{min} in the second cabling stage are approximated from the in-plane winding angle $\alpha_{\mathrm{in\text{-}plane}}$, the width w_{subcable} and the number n_{subcable} of RACC subcables using **equation 6.1**.

$$\tau_{\mathrm{min}} \approx w_{\mathrm{subcable}} \cdot n_{\mathrm{subcable}} \cdot \frac{1}{\cos\left(\alpha_{\mathrm{in\text{-}plane}}\right)} \qquad [6.1]$$

As shown in **table 6.4**, the CCRC concept is highly scalable. With the proposed geometry, 111 RACC subcables (4 mm wide) are possible at twist pitches τ comparable to ITER. Because of this, the twist pitch is not the limiting criteria in the scaling of CCRCs. The number of RACC subcables n_{subcable} which are transposed with the Rutherford cabling method increases the area necessary for the CCRC A_{CCRC}. This area A_{CCRC} can be approximated for round CCRCs (utilizing n_{subcable} 4 mm wide RACC subcables, 10 mm high) using **equation 6.2** and is the limiting factor in the winding packs of fusion magnets.

$$A_{\mathrm{CCRC}} \approx \left(\frac{n_{\mathrm{subcable}} \cdot 4\,\mathrm{mm}}{2\pi} + 10\,\mathrm{mm} \right)^{2} \cdot \pi \qquad [6.2]$$

---

[c] overall degradation = RACC degradation (3 %) $\cdot$ CCRC degradation (10 %) for the proposed geometry. The degradation is influenced by the in-plane winding angle $\alpha_{\mathrm{in\text{-}plane}}$ and the out-of-plane bending radius $r_{\mathrm{out\text{-}plane}}$ of the CCRC.

Table 6.4: Scaling of CCRCs to fusion relevant currents (68 kA at 12 T $\perp$) at cable temperatures of 4.2 K, 22 K and 50 K. RACC subcables with 100 strands (2 mm wide) corresponding to an effective *RE*BCO width per subcable of 0.2 m.

cable temperature T	4.2 K	22 K	50 K
J_c(77 K, self-field)	250 A/(cm−width)	250 A/(cm−width)	250 A/(cm−width)
lift factor: $J_c(T, 12\,\mathrm{T}\perp)/J_c$(77 K, self-field)	1.70	0.83	0.23
cabling degradation	13.3 %	13.3 %	13.3 %
$J_e(T, 12\,\mathrm{T}\perp)$	368.5 A/(cm−width)	179.9 A/(cm−width)	49.9 A/(cm−width)
effective *RE*BCO width	1.815 m	3.718 m	13.596 m
number of RACC subcables	9.1	18.6	68.0
cross sectional area $A_{\mathrm{CCRC}}{}^{\alpha}$	7.8 cm^2	15.2 cm^2	89.1 cm^2
minimal twist pitch $\tau_{\min}$	0.039 m	0.080 m	0.290 m
fraction of ITER twist pitch (0.42 m $\pm$ 0.02 m)	9.3 %	19.0 %	69.0 %

α approximation for round CCRCs with 4 mm wide RACC subcables (200 strands each), using **equation 6.2**.

6.3.5 Contacts

The CCRC concept utilizes small RACC subcables as strands. Thus, RACC contacting methods can be used in CCRCs but on a smaller scale and several times for one CCRC. Contacts of RACC cables are described in detail in **section 5.4**. Due to the lower number of *RE*BCO tapes per RACC subcable, the "stair" or "double stair" geometries are promising solutions. To homogenize the contact resistance between the RACC subcables, symmetric arrangements, with identical copper cross sectional areas and distances are necessary.

6.4 Conductor on Round Core (CORC) cable

Conductor on Round Core (CORC) cables are an HTS cable concept with an arrangement of the *RE*BCO tapes similar to *RE*BCO or BSCCO power cables. The cable concept has been proposed and published by D.C. van der Laan et al. in 2009 [Laa09] and is currently being developed by Advanced Conductor Technologies LLC[d].

6.4.1 Composition

Conductor on Round Core cables consist of *RE*BCO tapes which are tightly wound around a round former in several layers using a winding angle of 45°. The winding directions between

---

[d]Advanced Conductor Technologies LLC: http://advancedconductor.com/

each layer are reversed. In contrast to HTS power cables, formers with much smaller diameters are used. Existing CORC cables either utilize a copper rod or a copper power cable of 5.5 mm diameter as a former [LLG11]. In this geometry there are only three 4 mm wide tapes per layer in the three inner layers. In subsequent layers, the number of tapes increases due to the increase in winding diameter. The former provides electrical and mechanical stabilization. For additional mechanical stabilization the cable can be fitted with a jacket of structural material. Hollow formers, allowing forced flow cooling of the cable, are also possible. This cable concept is shown in schematic drawing in **figure 6.15**.

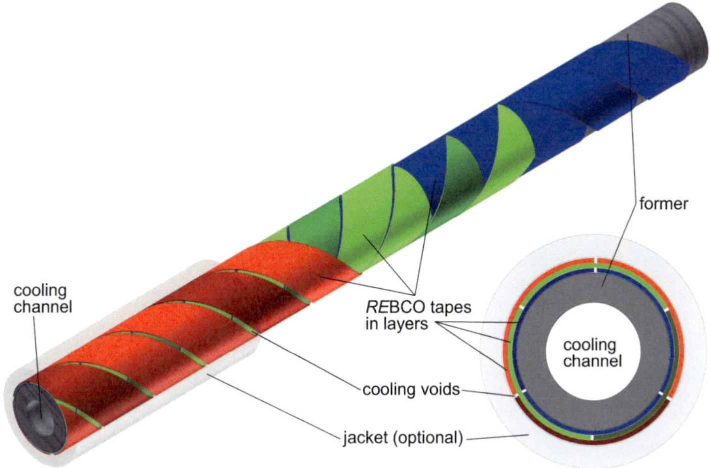

Figure 6.15: Schematic drawing of a Conductor on Round Core (CORC) cable. *RE*BCO tapes are tightly wound around a round former with a small diameter. The winding direction between the layers is reversed. The cable can optionally be fitted with a jacket of structural material for increased mechanical stabilization and with a hollow former for forced flow cooling. Picture after [Laa09].

Even though the tapes are wound around the former in different winding directions, the transposition of the *RE*BCO tapes in the CORC cable concept is only partial. For example, tapes of the inner layer always remain near the center, while tapes of the outer layer stay on the CORC cable's surface. The transposition remains partial regardless of whether the self-field of the cable, or just magnetic background fields are taken into account. The partial transposition may preclude the applicability of CORC cables as conductors in fusion magnets, and needs further investigation (e.g. AC-losses).

6.4.2 Mechanical properties

Due to the round geometry, CORC cables are isotropic in radial directions. Thus, in investigations of the mechanical properties of CORC cables, there are only two main load directions which have to be considered:

Mechanical loads parallel to the cable

In the conductors of magnets, this is the direction of the hoop stresses. The tapes of CORC cables are wound around the former in a winding angle of 45°. As shown in the appendix in **section A.1**, the influence of mechanical stress and strain on the current carrying capabilities of *RE*BCO tapes is highly anisotropic. CORC cables are wound with an angle of 45°. At this angle, winding stresses and hoop stresses, act on the a-axis (100 direction) and b-axis (010 direction) of all *RE*BCO crystal unit cells to the same extent, resulting in the lowest strain effect [LLG13]. Thus, CORC cables are in longitudinal direction even less sensitive to mechanical loads than single *RE*BCO tapes. At the critical stress limit[e], the current carrying capabilities of CORC cables are reduced by 3 % compared with a 10 % reduction in single straight *RE*BCO tapes [Laa12]. Because of this, CORC cables are ideal conductors in applications with high mechanical loads in longitudinal direction, e.g. the hoop stresses of magnets.

Radial mechanical loads

In magnets, the Lorentz forces are perpendicular to the conductor. Due to the radial symmetry of CORC cables, there are no preferred directions. Loads in the radial direction result in compressive transversal stress on the *RE*BCO tapes, perpendicular to the tape surface. In this direction, the tapes themselves are able to withstand more than 240 MPa pressure (SuperPower tapes, see **subsection 2.3.3.2**). Thus, CORC cables are expected to be insensitive to radial mechanical loads, too. The radial load direction is investigated through critical current measurements in magnetic background fields up to 12 T in **subsection 6.4.3**.

6.4.3 Field and temperature dependent measurements

A CORC cable sample is provided by Advanced Conductor Technologies LLC (D. van der Laan) for magnetic field and temperature dependent tests. The following experiments are conducted in cooperation with D. van der Laan[f]. The sample parameters are given in **table 6.5**.

The current carrying capabilities of the sample are first determined in a liquid nitrogen bath (77 K) in self-field conditions. After that, the cable is installed into the temperature variable insert of the FBI test facility. It is characterized at different surface temperatures in magnetic background fields up to 12 T. The magnetic field is increased and decreased to check the cable for damage due to the Lorentz loads at high magnetic fields.

---

[e]CORC cable consisting of 4 *RE*BCO tapes and 2 dummy tapes wound around a solid copper former in 2 layers. Due to stress concentration at the conic copper terminations, initial failure of strands at 144 MPa. However, the critical stress limit of the cable is expected to be significantly higher [Laa12].

[f]Affiliation: Advanced Conductor Technologies, USA and University of Colorado, USA

Table 6.5: Parameters of the tested CORC cable sample.

parameter	CORC cable sample
sample length	1.16 m including terminations
superconductor	4 mm wide copper stabilized from SuperPower (SCS4050) - no advanced pinning
number of tapes	15 tapes in 5 layers (3 tapes per layer)
twist pitch τ	17 mm
termination	all tapes individually soldered to cone shaped copper contacts
mechanical stabilization	power cable with 5.5 mm diameter as former
electrical stabilization	copper stabilization of *REBCO* tapes + copper in the strands of the power cable
voltage taps	3 taps at tapes in the outer layer + 1 tap at the terminations

6.4.3.1 LN$_2$ characterization

The transition of the CORC cable from superconducting state to normal conduction is abrupt at 77 K, self-field. Thus, a $1\,\mu\mathrm{V\,cm}^{-1}$ criteria is used to determine the critical currents in all the following experiments. The current carrying capabilities at these boundary conditions are shown in **table 6.6**. Ohmic contributions at the copper terminations (also referred to as the contacts) are subtracted. The n-values of the transitions cannot given, as the sample is not measured up to $10 \cdot E_\mathrm{c}$ ($10\,\mu\mathrm{V\,cm}^{-1}$). To protect the sample, the current is shut down just above the critical electric field.

Table 6.6: 77 K, self-field current carrying of a CORC cable consisting of 15 tapes (SuperPower, 4 mm wide, no advanced pinning). Voltage taps are attached to the tapes in the outer layer (tap1, tap2 and tap3) and at the copper terminations (contacts). Ohmic contributions at the copper terminations are subtracted. A $1\,\mu\mathrm{V\,cm}^{-1}$ criteria is used to determine the critical currents.

voltage tap:	tap1	tap2	tap3	contacts
I_c:	1752 A	1652 A	1757 A	1638 A

Voltage taps are only available at 3 of the 15 *REBCO* tapes of the sample. Even these tapes transition at slightly different currents (1652 - 1757 A). Thus, the voltage drop across the copper terminations, the contacts, is used to determine the current carrying capabilities of the cable. This voltage corresponds to an averaged behavior of all tapes. At 77 K, self-field the critical current of this CORC cable therefore is $I_\mathrm{c}\,(77\,\mathrm{K}, \mathrm{self\text{-}field}) = 1638\,\mathrm{A}$ (see [BDF12a, p. 7]). In all the following experiments, the voltage drop across the copper terminations (contacts) is used for the determination of the current carrying capabilities as well.

6.4.3.2 In-field measurements

After the LN$_2$ self-field testing, the current carrying capabilities of the sample are characterized with the FBI test facility at different magnetic field and temperatures. The sample and the temperature variable insert are shown in **figure 6.16**.

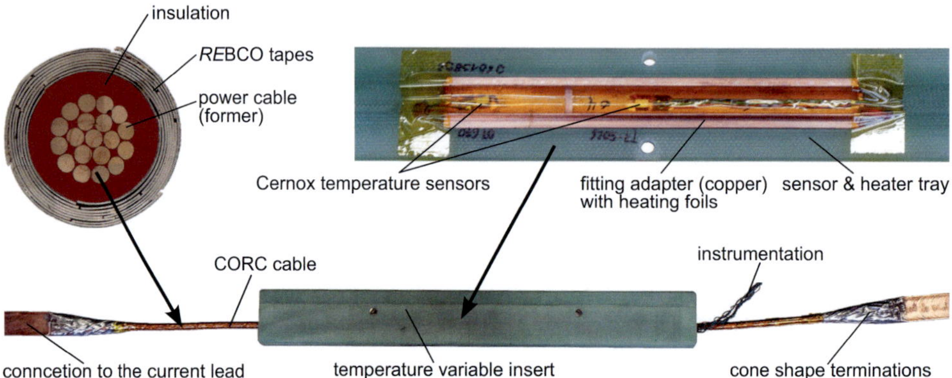

Figure 6.16: CORC sample in the temperature variable insert for field and temperature dependent measurements. Cross section view picture from [LLG11].

With a twist pitch of 17 mm, slightly more than 4 transposition lengths of the cable are inside the 70 mm long high field region of the FBI test facility. In this region the magnetic field deviation is less than 3 % from its peak value.

In the temperature variable insert, 100 mm long heating foils completely surround the sample. The heated section of the sample and the high field region magnet are aligned. The temperature of the cable is highest in the 70 mm long high field region, too. This region limits the current which can be carried by the cable and determines the critical current. Because of this, a distance of 70 mm is used to calculate the electric field in all field and temperature dependent measurements regardless of the separation of the voltage taps.

The magnetic background field is increased from zero field to 12 T in 2 T steps. Using the temperature variable insert, the current carrying capabilities of the CORC cable are measured for different temperatures in each step. Regardless of field or temperature, the shape of the transitions from superconducting state to normal conduction remain similar. All transitions are swift, allowing the determination of the critical current with the $1\,\mu V\,cm^{-1}$ criteria. In **figure 6.17**, two electric field vs. current curves are shown: high temperature (on the left) and high magnetic background field (on the right).

The shapes of these curves are exemplary for all superconducting transitions of this sample, implying constant performance in the whole magnetic field and temperature range.

With the copper in the *REBCO* tapes and the additional copper of the former, the electrical stabilization of the sample is sufficient. No significant increase of the temperature is observed

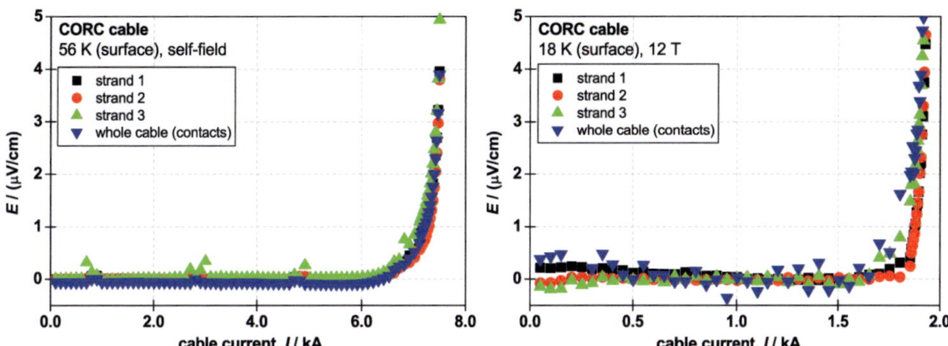

Figure 6.17: Example electric field vs. current curves for the superconducting transitions in the field and temperature dependent measurement of a CORC cable: high surface temperature (left) and high magnetic background field (right).

during superconducting transitions even close to the maximal current of the test facility. No degradation of the current carrying capabilities of the CORC cable samples is observed on field increases or decreases. The performance of the cable remains constant during all tests.

As shown in **figure 6.18**, critical currents are measured for sample surface temperatures from 4.2 K to 38.3 K in the whole magnetic field range using the $1\,\mu V\,cm^{-1}$ criteria. The voltage taps at the copper contacts are used to average the behavior of all tapes. Ohmic copper contributions are subtracted.

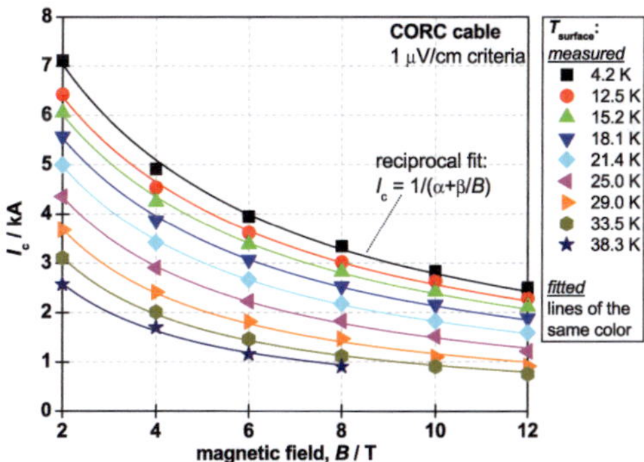

Figure 6.18: Field and surface temperature dependent measurements of a CORC cable (15 *REBCO* tapes in 5 layers). The measurements (points) use the voltage taps at the copper terminations (contacts). Ohmic contributions are subtracted. Critical currents are determined with the $1\,\mu V\,cm^{-1}$ criteria. Data is fitted with reciprocal functions (lines).

The curves of different sample surface temperatures are regularly spaced. The reciprocal function ($I_c = {}^1\!/(\alpha + \beta \cdot B)$) can be used to describe the reduction of the current carrying capabilities with increasing magnetic background fields. For the investigated temperature and magnetic field range, these functions fit the measured data quite well. The fitting parameters are given in the **appendix A.4** in **table A.1**. With increasing surface temperatures, the fitting parameters of the reciprocal functions increase. At 4.2 K, 12 T $\perp$ the cable carries 2.53 kA corresponding to a current density of 422 A/cm−width. This perfectly matches single tape data published by the *REBCO* tape manufacturer (single SuperPower tapes: 425 A/cm−width at 4.2 K, 12 T $\perp$ [Supd][Haz12, p. 8]). Thus, the current carrying capabilities of the *REBCO* tapes are fully utilized in the CORC cable concept.

6.4.4 High current ramping rates

All measurements of HTS cables are done with low current ramp rates. The current is always increased in small steps. Between the steps, the current is kept constant while the voltages of the tapes and the copper terminations are measured using low noise nano voltmeters. In magnets however, the current of the conductors can change swiftly during charging or fast discharging. An additional experiment is performed with the CORC cable sample. Instead of increasing the current in several steps, it is increased as fast as possible. The power supply of the FBI test facility provides a maximal current ramp rate of $8.3\,\text{kA s}^{-1}$. Using a low noise signal amplifier[g] and a fast data acquisition system[h], the voltage across the contacts of the CORC cable is measured during the ramping of the current. The measurement is done at self-field conditions and a cable temperature of 4.2 K. The data is shown in **figure 6.19**.

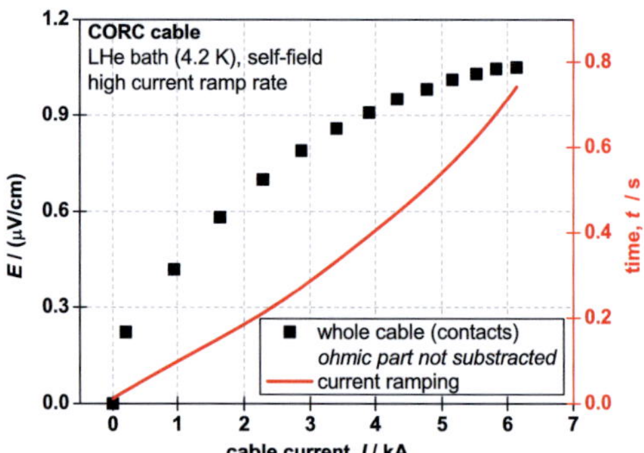

Figure 6.19: High current ramp rates on CORC cables. Current of the sample (4.2 K, self-field conditions) is increased at a ramp rate of $8.3\,\text{kA s}^{-1}$. The voltage drop between the copper terminations is measured. The ohmic contributions of the copper are not subtracted.

No excessive increase of the cable voltage is observed. Without the ohmic contributions of the copper terminations, the electric field remains below the critical electric field, even at the current ramp rates of $8.3\,\text{kA s}^{-1}$. This means that CORC cables can be expected to withstand fast changes of the current.

---

[g]voltage amplifier with a 3 dB frequency of 300 Hz
[h]data acquisition rate of 100 Hz

6.4.5 Temperature distribution

The CORC cable is heated from the outside with heating foils and the copper terminations of the sample are immersed in the liquid helium remaining at 4.2 K regardless of the heating. After the response time of the temperature variable insert (see **subsection 6.1.2.2**), stable temperature gradients develop. There is an temperature gradient parallel to the cable direction. The temperature is highest in the center of the heating section and decreases towards the borders of the heating foils. However, there is also a radial temperature gradient. In the heating section, the surface of the cable is in direct contact with the heating foils. This results in a high surface temperature and lower temperatures in the center of the sample.

The thermal conductivity of the *RE*BCO tapes is highly anisotropic. It is several orders of magnitude lower perpendicular to the tape surface (see **subsection 4.3.4.2**) compared with parallel to the tape surface (see **subsection 4.3.4.1**). Thus, the layers of *RE*BCO tapes in CORC cables thermally shield the center of the cable. The radial temperature gradient is therefore strongly pronounced.

The temperature of the CORC cable is measured only on the surface as no temperature sensors can be placed inside the sample. To obtain the distribution of the temperature in the heated section, the CORC cable is simulated using the finite element method. Three dimensional models of the CORC cable and the temperature variable insert are developed in real dimensions in the commercial software package Comsol. Temperature and direction dependent thermal properties are implemented for all materials used in the sample and the insert. For the *RE*BCO tapes, the thermal conductivity data given in **figure 4.3.4.1** and **figure 4.3.4.2** are used for parallel and perpendicular direction, respectively. Thermal conductivity data from **figure 4.17** and [BGS13, BGJW12] is used for copper, G10 insulation and solder. In the model, temperature sources (the heating foils) and temperature sinks (the helium bath) are imposed as boundary conditions. For optimal resolution, different mesh configurations are used. The *RE*BCO tapes are netted with a mapped mesh allowing the placement of several nodes within the superconducting layer of each tape. A free tetradic mesh, with much lower node density, is used in the temperature variable insert. This simulation method is shown schematically in **figure 6.20**.

With this method the temperature distribution is calculated for different surface temperatures. In regularly spaced positions along the cable, the temperatures of all tapes are averaged. In **figure 6.21**, this averaged cable temperature T_{average} is shown at different positions in the heating section for surface temperatures from 10 K to 100 K.

In the area of the heating foils, corresponding to the high field area of the magnetic field, the average temperature of the CORC cable is relatively stable. This zone is referred to as the "high temperature area" of the temperature variable insert. At 100 K cable surface temperature, the deviations of the average temperature in this 70 mm long zone are less than 4.2 %. At lower surface temperatures, the peak in the thermal conductivity of *RE*BCO tapes in parallel direction

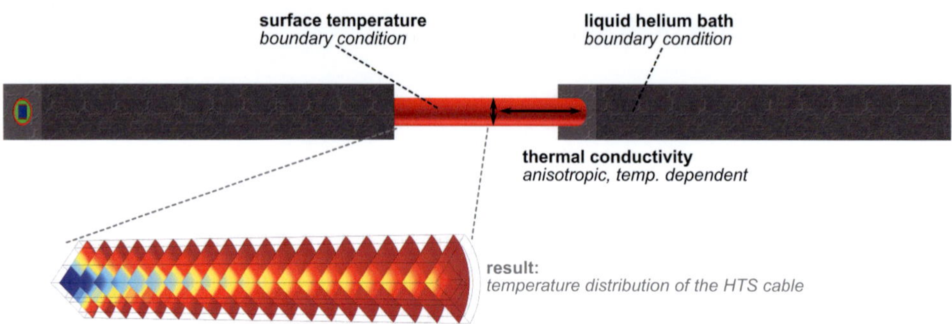

Figure 6.20: Schematic drawing of the FEM models used in the simulations of the temperature distribution of HTS cables in the temperature variable insert. An exemplary temperature distribution is shown as a color gradient from cold (blue) to hot (red).

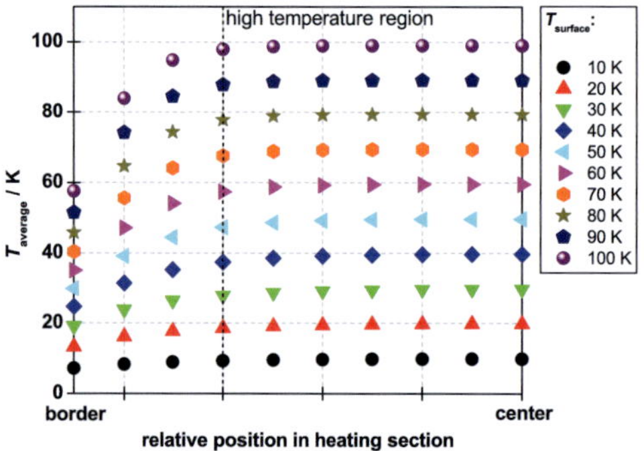

Figure 6.21: Simulated temperature distribution of a 15 tape CORC cable. The average temperature of the cable $T_{average}$ is shown at different positions along the cable (from the border to the center of the heating section).

(see **figure 4.16**) increases the thermal transport along the cable. At 38.3 K, which is the maximal surface temperature used in the field and temperature dependent characterization of the sample, the deviations of the average cable temperatures are less than 5.7 % in the high temperature region. With less than 6.2 % these deviations are maximal at 30 K cable surface temperature. At even lower surface temperatures, the deviations are reduced as the thermal conductivity of the *RE*BCO tapes becomes smaller. Additionally, lower surface temperatures mean lower temperature differences between the surface and the liquid helium bath, reducing the thermal transport in sample direction.

To summarize, the average temperatures of the cable in the inner, 70 mm long, section of the temperature variable insert are relatively constant. This zone corresponds to the high field region of the magnet. The deviations of the average cable temperature are less than 6.2 % for all investigated surface temperatures. A distance of 70 mm is therefore a good assumption to define the working area of the temperature variable insert.

The temperature dependence of the normalized critical current of the CORC cable sample is investigated at self-field using the temperature variable insert. The simulated temperature distribution of the sample (**figure 6.21**) is averaged in the 70 mm long high temperature region obtaining the average temperature of the sample $T_{average}$. The normalized critical current $I_c(T)/I_c(77\,\mathrm{K})$ depending on the average temperature $T_{average}$ is compared with single *REBCO* tapes published by the manufacturer [Haz12] in **figure 6.22**.

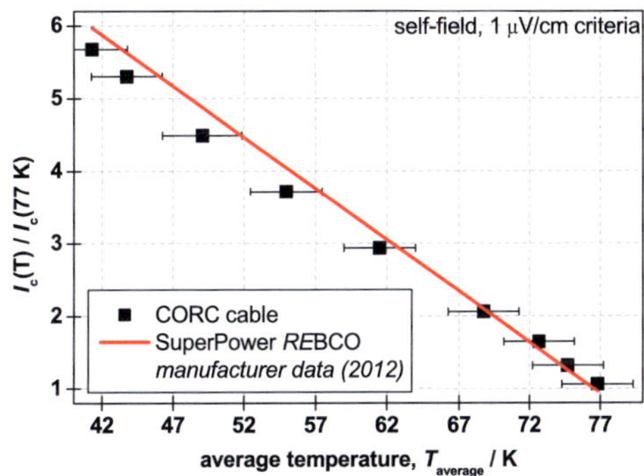

Figure 6.22: Temperature dependence of the normalized critical current of the CORC cable sample (black dots) and *REBCO* tapes from SuperPower (red line) at self-field conditions. The CORC cable is measured with the temperature variable insert of the FBI test facility. Its average temperature is calculated using the simulated temperature distribution show in **figure 6.21**. *REBCO* tapes data from [Haz12].

The CORC cable and *REBCO* tape manufacturer data are in good agreement within the measurement uncertainty. In the investigated temperature range of 40 - 77 K, the temperature dependence is linear for the tapes and for the CORC cable sample. The self-field critical current of the CORC cable sample is 1725 A using a $1\,\mu\mathrm{V\,cm^{-1}}$ criteria and the voltage taps at the copper terminations at 77 K average temperature $T_{average}$. In a liquid nitrogen bath, the self-field critical current of this sample is determined as 1638 A using the same voltage taps and critical electric field. With a deviation of 5.3 %, both self-field 77 K critical currents (temperature variable insert and liquid nitrogen bath) coincide and are within the 6.2 % temperature uncertainty of the temperature distribution of the temperature variable insert.

6.4.6 Scaling to currents relevant for fusion magnets

CORC cables can be scaled to very high currents by adding more layers of *REBCO* tapes. With increasing number of layers the diameter of the cable increases. At the same time, this increases the number of tapes per layer and also the twist pitch of the cable. The starting twist pitch of CORC cables of 17 mm is quite small, several hundreds of layers are possible within the stage 5 twist pitch of the ITER toroidal field coil conductors (0.42 m $\pm$ 0.02 m [ITE09e, p. 20]). The effective superconductor width necessary to carry fusion magnet relevant currents using CORC cables is calculated utilizing the same method and boundary conditions as for RACC cables (see **subsection 6.2.4.1**). However, as no degradation is observed in CORC cabling, the 3 % degradation factor is removed. This means that CORC cables require an effective *REBCO* width of 1.6 m (400 tapes 4 mm wide) at 4.2 K cable temperature. At cable temperatures of 22 K and 50 K, effective *REBCO* widths are necessary of 3.28 m (820 tapes 4 mm wide) and 12.0 m (3000 tapes 4 mm wide), respectively.

The scaling of CORC cables can be calculated iteratively. Starting with the former, the circumference of the cable is calculated for the diameter of the current layer. The circumference is divided by the product of the width of the *REBCO* tapes and the cosine of the winding angle. This quotient is rounded down resulting in the number of tapes in the current layer. The void area in the layer can be obtained from the difference of the circumference and the product of the number of tapes in the layer, the cosine of the winding angle and the width of the tapes. To the next layer, the diameter is increased by twice the thickness of the tapes. The total number of tapes and the total void area can be obtained by summation over all layers. With this iterative calculation method, the diameter of CORC cables, the resulting number of tapes and the total void area can be calculated depending on the number of layers. This calculation method is explained in detail within the appendix, **section A.5**. For CORC cables of up to 215 layers, the data is given in **table A.2**.

The diameter d (in mm) of CORC cables can also be approximated depending on the number of tapes n without resorting to an iterative method, using **equation 6.3**. A round former of 5.5 mm diameter, 4 mm wide SuperPower tapes and a winding angle of 45° are assumed.

$$d(n) = 3.27 \cdot 10^{-28} \cdot n^9 - 4.86 \cdot 10^{-24} \cdot n^8 + 3.07 \cdot 10^{-20} \cdot n^7 - 1.07 \cdot 10^{-16} \cdot n^6 + 2.26 \cdot 10^{-13} \cdot n^5$$
$$- 2.97 \cdot 10^{-10} \cdot n^4 + 2.46 \cdot 10^{-7} \cdot n^3 - 1.30 \cdot 10^{-4} \cdot n^2 + 5.88 \cdot 10^{-2} \cdot n + 5.73 \qquad [6.3]$$

Such an approximation is also possible for the dependence of the void area v (in mm^2) on the number of tapes n. Boundary conditions as before are assumed in **equation 6.4** for this calculation.

$$v(n) = 5.23 \cdot 10^{-28} \cdot n^9 - 7.71 \cdot 10^{-24} \cdot n^8 + 4.81 \cdot 10^{-20} \cdot n^7 - 1.66 \cdot 10^{-16} \cdot n^6 + 3.47 \cdot 10^{-13} \cdot n^5$$
$$- 4.52 \cdot 10^{-10} \cdot n^4 + 3.69 \cdot 10^{-7} \cdot n^3 - 1.92 \cdot 10^{-4} \cdot n^2 + 8.46 \cdot 10^{-2} \cdot n + 0.12 \qquad [6.4]$$

The approximated and iteratively calculated scaling behavior of CORC cables are compared in **figure 6.23**.

Iteratively calculated data (from **table A.2**) is shown as points. Approximated data using **equation 6.3** and **equation 6.4** is shown as lines. Both methods are in good agreement for the investigated range of tapes, making the 9th order polynomial functions an ideal method to

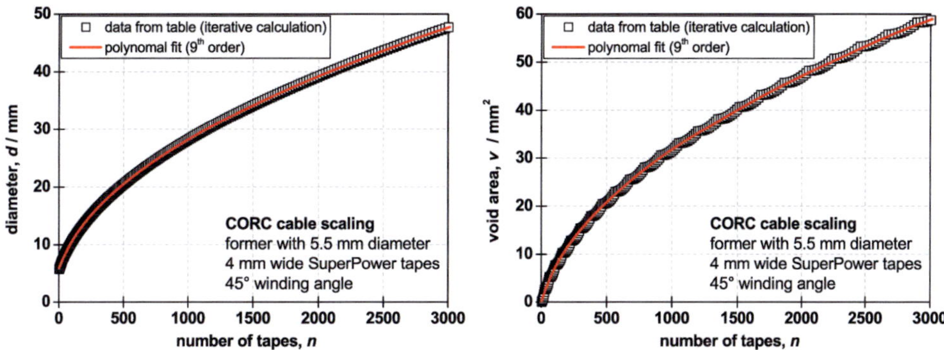

Figure 6.23: Scaling of CORC cables: Iteratively calculated (points), is compared with data fitted (lines) with 9th order polynomial functions. Iterative data is taken from **table A.2**. The polynomial fits use **equation 6.3** and **equation 6.4**. A round former of 5.5 mm diameter, 4 mm wide SuperPower tapes and a winding angle of 45° are assumed. Diameter vs. number of tapes is shown on the left and void area vs. number of tapes is shown on the right.

describe the scaling of CORC cables. Using these scaling laws, CORC cables with 4 mm wide tapes, a former of 5.5 mm diameter and a winding angle of 45° are scaled to carry fusion relevant currents (68 kA in perpendicular magnetic background fields of 12 T). This is shown in **table 6.7** for cable temperatures of 4.2 K, 22 K and 50 K.

Table 6.7: Scaling of CORC cables to fusion relevant currents (68 kA at 12 T $\perp$) at cable temperatures of 4.2 K, 22 K and 50 K. A former with 5.5 mm diameter and a winding angle of 45° are assumed.

cable temperature T	4.2 K	22 K	50 K
J_c(77 K, self-field)	250 A/(cm−width)	250 A/(cm−width)	250 A/(cm−width)
lift factor: $J_c(T,\ 12\,\text{T}\ \perp)/J_c$(77 K, self-field)	1.70	0.83	0.23
cabling degradation	0 %	0 %	0 %
$J_e(T,\ 12\,\text{T}\ \perp)$	425 A/(cm−width)	207.5 A/(cm−width)	57.5 A/(cm−width)
effective $REBCO$ width	1.815 m	3.718 m	13.596 m
number of 4 mm tapes	204	820	3002
number of layers	42	101	211
diameter d	13.9 mm	25.7 mm	47.7 mm
cross sectional area A_{CORC}	1.5 cm^2	5.2 cm^2	17.9 cm^2
void area v	11.4 mm^2	28.2 mm^2	58.7 mm^2
minimal twist pitch τ_{min}	0.040 m	0.147 m	0.290 m
fraction of ITER twist pitch (0.42 m $\pm$ 0.02 m)	9.4 %	18.9 %	35.0 %

6.4.7 Contacts

CORC cables use conical copper terminations as contacts up to date. Starting with the *RE*BCO tapes of the inner layer of the CORC cable, the tapes are individually soldered to close to the tip of the cone. All layers of tapes have to be processed one after another. With each layer, the number of tapes per layer can increase. However, the circumference of the cone and the available area increases, too. This method gives identical contact areas and conditions during soldering for all tapes. Homogeneous contact resistance in the 430 - 800 nΩ range is achieved for all tapes [Laa12, p. 45]. The size of the cone shaped contacts is scaled according to the number of tapes in the CORC cable. However, at several hundreds of tapes, the cones have to be quite large. Because of this, other contact concepts need to be investigated as well.

6.5 Twisted Stacked-Tape Cable (TSTC)

The Twisted Stacked-Tape Cable (TSTC) concept was proposed by M. Takayasu et al. in 2011 [TCBM11] to provide "simple, high current density and scalable cabling method applicable to a large scale magnet" [TMB11, p. 1]. The TSTC concept is currently developed at the Plasma Science and Fusion Center of the Massachusetts Institute of Technology (MIT)[i].

6.5.1 Composition

In Twisted Stacked-Tape Cables, several *RE*BCO tapes are stacked and twisted. 4 mm wide tapes are commonly used. 1 mm thick copper tapes of the same width can be attached to the top and the bottom of the *RE*BCO tape stack. The copper tapes provide additional electrical and mechanical stabilization. The twisted stack is inserted into a tube (sheath) of structural material. To prevent movement of the tapes and to avoid stress concentrations, all the voids between the TSTC and the sheath have to be filled. Glue, resin or solder are suitable filling materials. In **figure 6.24**, a single TSTC is shown schematically.

The number of tapes in a single TSTC stack is limited. A thickness of the stack equal to the width of the tapes is ideal. For 4 mm wide SuperPower tapes this means stack sizes of around 40 tapes. To further increase the current carrying capabilities, several TSTCs have to be combined. Similar to the RACC cables in the CCRC concept, the TSTC stacks can be subcables of a larger cable. Several TSTC stacks can be inserted into grooves of a former of structural or stabilizing material to stabilize the cable. For additional mechanical stabilization, such multi-stack TSTCs can be also equipped with jackets of structural materials. A three stack TSTC is shown in a schematic drawing in **figure 6.25**.

---

[i]Plasma Science and Fusion Center: http://www.psfc.mit.edu/

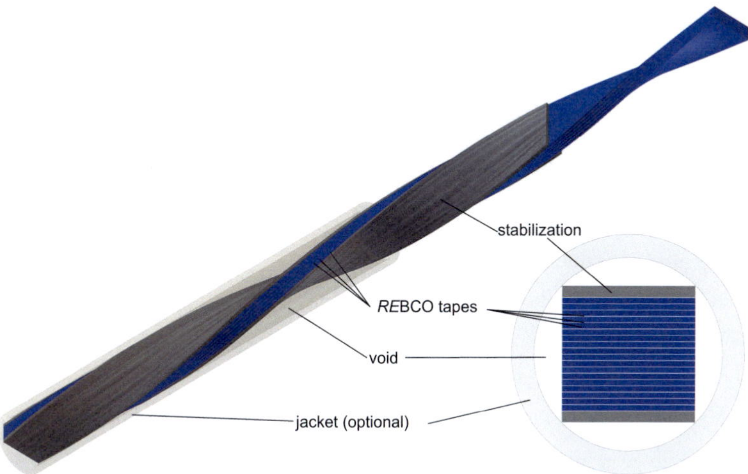

Figure 6.24: Schematic drawing of a single Twisted Stacked-Tape Cable (TSTC). *RE*BCO tapes are stacked and twisted. The mechanical and electrical stabilization is increased with copper tapes on the top and the bottom of the superconductor stack. To further improve the mechanical properties, the TSTC can be equipped with a sheath of structural material.

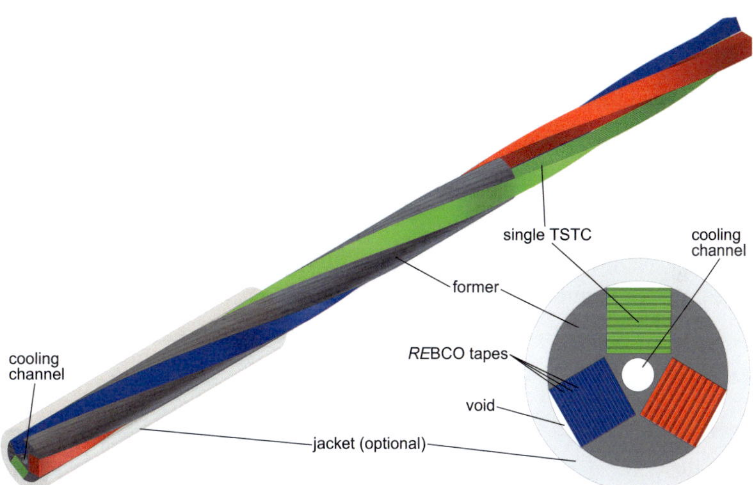

Figure 6.25: Schematic drawing of a TSTC consisting of three stacks. Rutherford cabling is used, fully transposing the *RE*BCO stacks. For mechanical stabilization, the stacks are inserted into grooves of a former. A jacket of structural material can further enhance the mechanical properties. Picture after [MBT11, p. 21].

In the stack, the *RE*BCO tapes are not transposed completely even if the stack is twisted. Inner tapes always remain in the center of the stack, while outer tapes remain on the outside. The twisting of the stacks can be considered as the first stage transposition of the whole TSTC. In TSTCs consisting of several stacks, there is also a second stage transposition, the transposition of the stacks themselves. The Rutherford cabling of the second stage, completely transposes all the twisted superconductor stacks. Therefore, two scenarios have to be considered.

Considering self-fields

As the *RE*BCO tapes are not completely transposed in the first stage of cabling, their overall transposition remains partial [Bru13a].

Considering only magnetic background fields

For magnetic background fields, the partial transposition of the first cabling stage does not matter. The transposition of the superconductor stacks themselves in the second stage of cabling is sufficient to call the overall transposition of TSTCs complete [Bru13b].

Due to the high magnetic background fields in the conductors of fusion magnets, the self-field of each cable is close to negligible. Because of this, TSTCs qualify as fully transposed conductors in fusion magnets.

6.5.2 Mechanical properties

Due to their round geometry, TSTCs are isotropic in radial directions. Because of this, there are only two main load directions to be considered:

Mechanical loads parallel to the cable

This is the direction of the hoop stresses in magnet conductors. Due to the nearly straight arrangement of the *RE*BCO tapes in the TSTC concept, the mechanical performance of TSTCs is determined in the parallel direction by the average mechanical properties of the former and the tapes. With the high critical tensile stress limit of *RE*BCO tapes of 500 - 700 MPa (see **subsection 2.3.3.2**), similar stress limits can be expected for whole TSTCs if stainless steel formers are used. The TSTC concept is a good choice in applications with high mechanical loads in the parallel direction, e.g. high field magnets.

Radial mechanical loads

In magnets, there are also components of the forces perpendicular to the conductors, e.g. the radial components of the Lorentz forces. Similar to CORC cables, there are no preferred perpendicular directions in TSTCs because of their radial symmetry. Applying perpendicular forces of any direction (in the following referred to as radial forces), presses the *RE*BCO tapes of single stack or multi-stack TSTCs against the jacket or the former, respectively. Due to the twisting of the stacks, the contact points of the loads repeatedly shift between the plains and the edges. The edges of the superconductor stacks are prone to stress concentrations. Such stress concentrations are especially critical in single stack TSTCs. In the single stack cables, concentration of stress at the edges can be avoided by filling the voids. This connects the stack and the sheath, prevents movement of the outer tapes and evenly disperses the mechanical loads.

In multi-stack TSTCs the former helps to distribute the radial loads. The influence of radial mechanical loads on the current carrying capabilities of single stack TSTCs is investigated through critical current measurements in high magnetic background fields in **subsection 6.5.3**.

6.5.3 Field and temperature dependent measurements

A TSTC sample with one superconductor stack is provided by the Massachusetts Institute of Technology for magnetic field and temperature dependent tests. The experiments are performed together with M. Takayasu[j] at the Institute for Technical Physics of the Karlsruhe Institute of Technology. The parameters of the sample are shown in **table 6.8**.

Table 6.8: Parameters of the tested TSTC sample.

parameter	TSTC sample
sample length	1.16 m including terminations
superconductor	4 mm wide copper stabilized from SuperPower (SCS4050) - advanced pinning
number of tapes	40
twist pitch τ	200 mm
termination	clamped BSCCO - $REBCO$ connections
mechanical stabilization	twisted stack soldered into Cu tube (7.90 mm diameter, 0.81 mm wall thickness)
electrical stabilization	Cu stabilization of $REBCO$ tapes + Cu tapes on top and bottom of the stack
voltage taps	3 taps on the copper tube + 1 tap at the terminations

As the superconductor stacks are soldered into a copper tube, the TSTC sample qualifies as a cable in conduit conductor (CICC). The hollow space between the stack and the tube is completely filled with soft solder preventing movement of the stack and dispersing mechanical loads to avoid stress concentrations near the stack's edges. The sample is installed in the temperature variable insert of the FBI test facility using fitting adapters for round samples and is characterized at different surface temperatures and magnetic background fields. The characterization is done in two steps:

- cycling of the magnetic background field and the radial mechanical load on the sample (**subsection 6.5.3.1**).

- critical current measurements at different fields and temperatures (**subsection 6.5.3.2**).

---

[j]Affiliation: Massachusetts Institute of Technology, USA

6.5.3.1 Critical current measurements at increasing and decreasing magnetic fields

Initially, the cable is checked for degradation of the current carrying capabilities due to high radial mechanical loads. This is achieved through cycling of the magnetic background field, currents and radial loads. The magnetic background field is increased from zero field to 12 T. Every 2 T, the critical current is measured at 4.2 K.

In the 8 - 12 T range, the current carrying capabilities of SuperPower advanced pinning *REBCO* tapes are only slightly reduced with magnetic field increases. This is visible as a shallow curve in **figure 2.7**. Consequently, the Lorentz forces, and the radial loads on the cable, are maximal at 12 T. After reaching maximal field, and maximal radial loads, the background field is reduced back to zero. Once again, the critical currents are measured every 2 T. Degradation of the current carrying capabilities is possible, due to the enormous mechanical loads. The degradation is exposed by comparing the critical currents measured at identical background fields on field increases and decreases. This procedure is repeated several times to detect further degradation of the sample. The measured critical currents are shown in **figure 6.26**. In these measurements, the sample is kept at 4.2 K, the temperature variable insert is not activated. The voltage taps at the copper contacts are used to average over the behavior of all the tapes. The ohmic contributions of the copper in the sheath and in the terminations are subtracted. The superconducting transitions of the TSTC are not as steep as the transitions of the CORC cable sample from **subsection 6.4.3**. Thus the n-values of the TSTC are lower, as well. A $5\,\mu\text{V}\,\text{cm}^{-1}$ criteria is therefore used to determine the critical currents in all measurements of the TSTC sample.

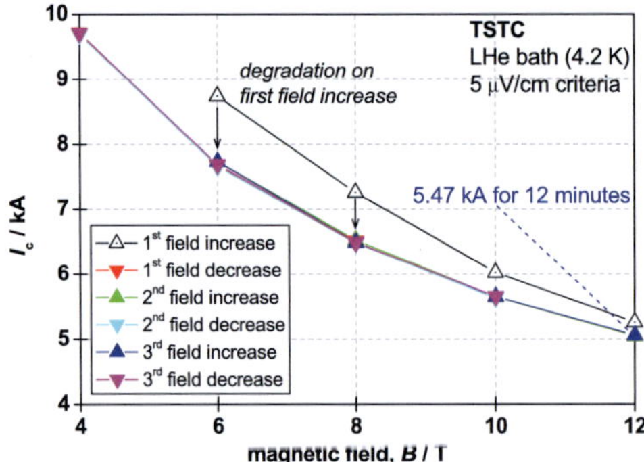

Figure 6.26: Critical current measurements at increasing and decreasing magnetic background fields of a 40 tape TSTC. Voltage taps at the terminations are used, the ohmic contributions of the copper in the sheath and in the terminations are subtracted.

On the first increase of the magnetic background field, the current carrying capabilities of the sample degrade. However, the degradation does not continue during following magnetic field cycles. After the first cycle, the current carrying capabilities remain unchanged even if the TSTC is kept at the maximum current of 5.47 kA at 12 T for 12 min. The critical currents measured on the first increase of the magnetic field and on later field cycles are shown in **table 6.9**. The current carrying capabilities of all magnetic field cycles but the first field increase are averaged.

Table 6.9: Degradation of a 40 tape TSTC due to radial mechanical loads at high magnetic background fields. Critical currents of field cycles, following on the first magnetic field increase, are averaged.

field	critical current - first increase	critical current - later	degradation
6 T	8.74 kA	7.70 kA	11.9 %
8 T	7.26 kA	6.50 kA	10.5 %
10 T	6.02 kA	5.64 kA	6.3 %
12 T	5.26 kA	5.05 kA	4.0 %

Determination of degradation is not possible at fields of 4 T and below as the critical currents of the sample are above the maximal current of the test facility (10 kA). At background fields of 6 T, a maximum of the degradation of 11.9 % is obtained. For higher fields, the degradation decreases. The non-constant degradation implies that damage to the sample occurs not just during the critical current measurement at maximal field. Some degradation already occurs below maximal magnetic fields. The factor of degradation is nearly identical at 6 T and 8 T, signifying that these fields are safe, and that the sample is damaged during the critical current measurements at 10 T and 12 T magnetic background field, only.

Radial mechanical loads can damage the *REB*CO tapes in TSTCs, as pointed out by these degradations. In the investigated sample, the voids between the stack of superconductor tapes and the jacket are filled with soft solder. Due to its weak mechanical properties and challenging filling procedure, soft solder is not an ideal filler. In the investigated TSTC, the solder was heated, liquified and filled through holes into the copper tube, which is placed on a heater. With this method there may be some areas in which the solder did not penetrate completely. During the first field increase, the Lorentz forces probably broke the stack of the *REB*CO tapes loose in some regions. This leads to a concentration of mechanical stresses and damages the tapes. A mixture of Araldite epoxy resin and quartz powder of a mixing ratio of 1:1 is a much better choice as filler. As shown in **section 4.2.3**, this mixture can be used to impregnate HTS cables without causing degradation of their current carrying capabilities. In contrast to solder, epoxy resin mixtures can be drawn easily into the sample through repeated vacuum impregnation. The degradations on the first field increase may have been prevented by filling the TSTC with a epoxy resin mixture instead of solder.

6.5.3.2 Temperature dependence at different background fields

In a second step, the magnetic background field is increased for the final experiment. In contrast to the magnetic field cycles, the temperature variable insert is now activated. The critical currents are measured in each 2 T step for different cable temperatures. If possible, the same temperature steps are used over the whole magnetic field range.

The two copper tapes on the top and the bottom of the *REBCO* tapes stack and the encasing copper tube electrically stabilize the sample. The stabilization is sufficient, there is no thermal runaway, even during superconducting transitions close to 10 kA transport current.

In **figure 6.27**, the current carrying capabilities of the TSTC sample are shown for sample surface temperatures from 4.2 K to 37.8 K using a $5\,\mu V\,cm^{-1}$ criteria. As in the previous measurements, the voltage taps at the copper contacts are used. Ohmic contributions are subtracted.

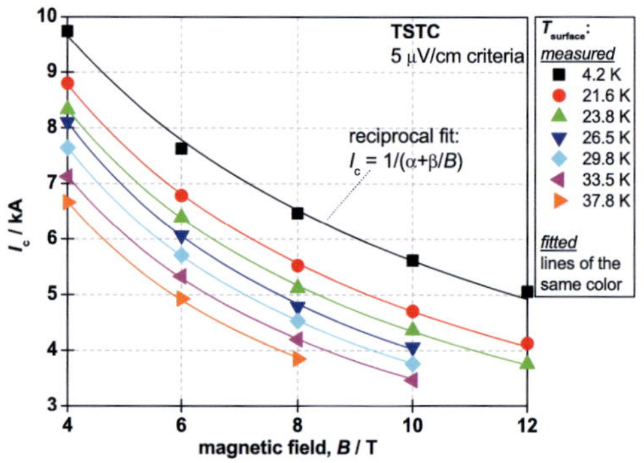

Figure 6.27: Field and surface temperature dependent measurements of a 40 tape TSTC. The measurements (points) use the voltage taps at the copper terminations (contacts). All ohmic contributions are subtracted. Critical currents are determined with the $5\,\mu V\,cm^{-1}$ criteria. Data is fitted with reciprocal functions (lines).

The curves of different surface temperatures are regularly spaced. Similar to the investigation of the CORC cable in **section 6.4.3**, reciprocal functions ($I_c = 1/(\alpha+\beta \cdot B)$) are used to describe the critical current and magnetic field dependence. Measured and fitted data match well in the investigated field and temperature range. The fitting parameters of the reciprocal functions are given in the **appendix A.6** in **table A.3**. Similar to the CORC cable, the slope of the critical current vs. magnetic background field curves is reduced with increasing sample surface temperatures.

The n-values are calculated for all superconducting transitions of the field and temperature dependent measurements. The calculations are done in the electric field range from $1\,\mu V\,cm^{-1}$ to $10\,\mu V\,cm^{-1}$. The voltage taps at the copper contacts are used. Once again, the ohmic contributions are subtracted. The n-values are shown in **figure 6.28**.

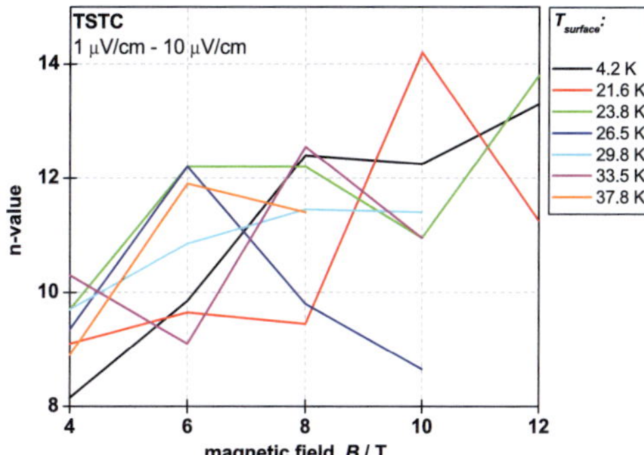

Figure 6.28: The n-values of the superconducting transitions of a 40 tape TSTC. The measurements are of the degraded cable at different cable temperatures and magnetic background fields. The n-values are calculated in the electric field range 1 - 10 μV cm^{-1} using the voltage tapes at the copper contacts. Ohmic contributions subtracted.

All n-values are relatively low, they are in the 8 to 14 range. There are two aspects contributing to the low n-values:

- Firstly, the degradation of the sample, which occurred on the first field increase. Some of the *RE*BCO tapes were damaged, reducing the current they can carry. This damage in-homogenizes the cable, causing prematurely transitions of some tapes at lower currents. Measured at the copper contacts of the sample, which give an average behavior of all tapes, the steepness of the superconducting transitions is reduced.

- Secondly, there is the copper jacket to consider. The superconductor stacks are soldered into a copper tube. Eddy currents can develop in the copper and generate voltages in addition to the electric field of the *RE*BCO tapes. This additional voltage reduces the steepness of the superconducting transitions and the n-values.

In a non-degraded TSTC, especially one without a copper jacket, significantly higher n-values are to be expected.

6.5.4 Temperature distribution

The distribution of the temperature in the heating section of the temperature variable insert is calculated analog to the CORC sample (as described in **subsection 6.4.5**). The geometry and the thermal properties of the constituents of the TSTC sample are used in the finite element method model.

The temperature distribution is calculated for surface temperatures from 10 - 100 K. The temperatures of all tapes are averaged at regularly spaced positions along the TSTC. These temperatures $T_{average}$ are shown **figure 6.29** from the border to the center of the heating section.

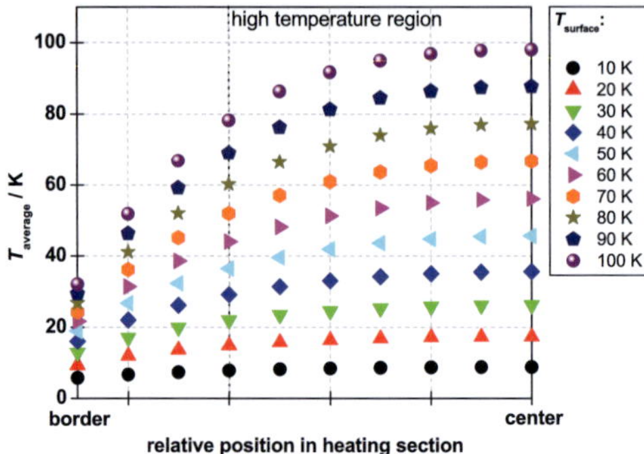

Figure 6.29: Simulated temperature distribution of a 40 tape TSTC. The average temperature of the cable $T_{average}$ is shown at different positions along the cable (from the border to the center of the heating section).

Compared with the CORC cable sample, the differences of the average cable temperatures $T_{average}$ are larger in the TSTC sample. With the copper sheath and the copper tapes on the top and the bottom of the superconductor stack, the TSTC contains a lot of highly conductive material. This strongly increases the thermal transport along the sample and leads to lower average temperatures near the borders of the heating section. In the CORC cable, the copper is contained in the strands of the central power cable, the former of the sample. The electrical insulation of the power cable thermally insulates its copper strands from the rest of the CORC cable. Because of this, the impact of the copper on the thermal transport is much higher in the TSTC sample compared with the CORC cable.

Even in the central 70 mm of the temperature variable insert, the "high temperature region", the deviations of the average $REBCO$ tape temperatures are significant. At 100 K cable surface temperature, the deviations are 20.2 %. In the surface temperature range of interest, 20 - 40 K, the deviations are with 18.2 % maximal at 40 K and go down to 14.4 % at 20 K surface temperature. From these strong deviations it is evident that the thermal transport along the TSTC sample is just too high for the temperature variable insert in its current layout. A significantly longer heating section and better thermal insulation outside the high temperature region is necessary for samples with high copper contents.

6.5.5 Scaling to currents relevant for fusion magnets

During the cabling process of the TSTC concept, the current carrying capabilities of the tapes are degraded by 3 % [TCBM11, p. 4]. The superconductor width necessary to carry fusion relevant currents (68 kA at 12 T $\perp$) is calculated with the same method and boundary conditions as for RACC cables (see **subsection 6.2.4.1**). A square cross sectional area of the superconductor stack is ideal. Because of this, a single TSTC stack consist of 40 copper stabilized SuperPower tapes,

4 mm wide each, corresponding to an effective *REBCO* width of 0.16 m. Several of these single stacks have to be combined into a larger TSTC to be usable as conductor in fusion magnets. The minimal twist pitch τ_{min} of multi-stack TSTCs can be approximated using **equation 6.5**, where α is the winding angle, $w_{subcable}$ is the edge length and $n_{subcable}$ the number of TSTC stacks.

$$\tau_{min} \approx w_{subcable} \cdot n_{subcable} \cdot \frac{\sqrt{2}}{\cos(\alpha)} \qquad [6.5]$$

The cross sectional area A_{TSTC} of these multi-stack TSTCs can be approximated using **equation 6.6**.

$$A_{TSTC} \approx \left(\frac{n_{subcable} \cdot 4\,mm}{2\pi} + 4\,mm \right)^2 \cdot \pi \qquad [6.6]$$

As shown in **table 6.10**, the required amounts of *REBCO* tape (effective width), the corresponding number of stacks and the minimal twist pitch τ_{min} are calculated for multi-stack TSTCs at cable temperatures of 4.2 K, 22 K and 50 K.

Table 6.10: Scaling of TSTC cables to fusion relevant currents (68 kA at 12 T $\perp$) at cable temperatures of 4.2 K, 22 K and 50 K. Winding angles of 20° and stacks with square cross sectional areas of 4 mm edge length consisting of 40 *REBCO* tapes (4 mm wide and 0.1 mm thick) are assumed.

cable temperature T	4.2 K	22 K	50 K
J_c(77 K, self-field)	250 A/(cm−width)	250 A/(cm−width)	250 A/(cm−width)
lift factor: $J_c(T, 12\,T\,\perp)/J_c$(77 K, self-field)	1.70	0.83	0.23
cabling degradation	3 %	3 %	3 %
$J_e(T, 12\,T\,\perp)$	412 A/(cm−width)	201 A/(cm−width)	55 A/(cm−width)
effective *REBCO* width	1.65 m	3.38 m	12.36 m
number of stacks	10.3	21.3	77.3
cross sectional area A_{TSTC}	3.5 cm^2	9.7 cm^2	88.9 cm^2
minimal twist pitch τ_{min}	0.066 m	0.132 m	0.470 m
fraction of ITER twist pitch (0.42 m $\pm$ 0.02 m)	15.8 %	31.5 %	111.8 %

By adding more *REBCO* tape stacks, the TSTC concept can be scaled up to fusion relevant currents even at elevated cable temperatures. At 50 K cable temperature, the twist pitch of ITER is exceeded. Comparable to the CCRC concept, the number of TSTC stacks strongly increases the necessary cross sectional area A_{TSTC}. The cross sectional area is the limiting factor in the winding packs of fusion magnets.

6.5.6 Contacts

For contacting TSTCs, a two step contacting method is available [TMB11]. In a first joint,
*REB*CO tapes of TSTC stacks are sandwiched with BSCCO tapes over long lengths. They are
either soldered or clamped. For clamping, contact pressures in the 55 MPa range [TMB11] are
necessary. Soldering results in lower contact resistances, while clamping allows the demounting
of the joints. In a second step, the BSCCO tapes are soldered into the copper termination. This
two step contacting method has advantages similar to the the "stair" shaped individual contacting
method for RACC cables. For all *REB*CO tapes, the contact conditions and contact areas are
identical, homogenizing the contact resistances. The two step contacting method is shown
schematically in **figure 6.30**.

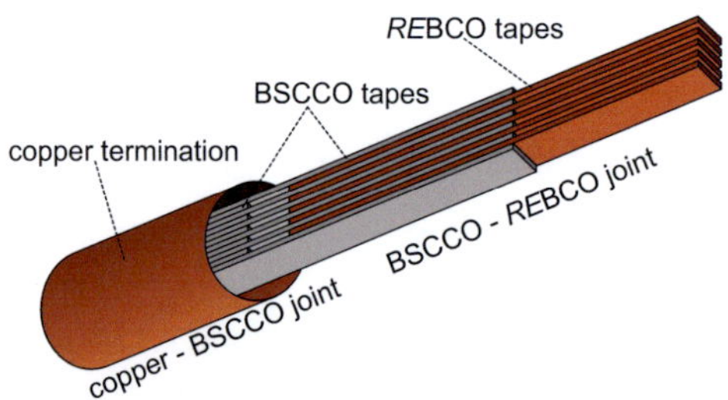

Figure 6.30: Schematic drawing of a two-step contacting method for TSTCs. In the first joint,
*REB*CO tapes are sandwiched with BSCCO tapes. In the second joint, the BSCCO tapes are
soldered into a copper termination. Picture after [TMB11].

Soldering BSCCO tapes into copper terminations works much better than with *REB*CO tapes.
Due to their silver matrix, the current can penetrate BSCCO tapes from all sides. Additionally,
this allows currents to pass from one tape into another inside the contacting area. This is not even
remotely possible with *REB*CO tapes, as the buffer layers and the substrate electrically insulate
the tapes from one side. Thus, *REB*CO tapes can only be contacted from their superconductor
side. Even in copper stabilized *REB*CO tapes, almost no current can pass from one side of the
tapes to the other.

By adding more BSCCO tapes, this contacting method can be scaled to higher *REB*CO tape
numbers. However, due to the strong decrease of the current carrying capabilities of BSCCO
tapes in high magnetic background fields (see **figure 2.7**), this contacting method has to be used
outside the high field area of magnets.

6.6 Round Strands Composed of Coated Conductor Tapes (RSCCCT) cable

The Round Strands Composed of Coated Conductor Tapes (RSCCCT) cable concept was proposed by D. Uglietti et al. in 2012 [UWB13b]. It is currently developed at the Centre de Recherches en Physique des Plasmas (CRPP) of the École Polytechnique Fédérale de Lausanne (EPFL)[k].

6.6.1 Composition

This HTS cable concept consists of two stages of cabling. In the first cabling stage, stacks of *RE*BCO tapes are soldered between two semicircular copper profiles. Each *RE*BCO stack contains only a few tapes, thus the resulting conductor is nearly circular. It is twisted, partially transposing the superconducting tapes. Twist pitches from 280 mm to 360 mm are used. The first stage of the Round Strands Composed of Coated Conductor Tapes cable concept is shown schematically in **figure 6.31**.

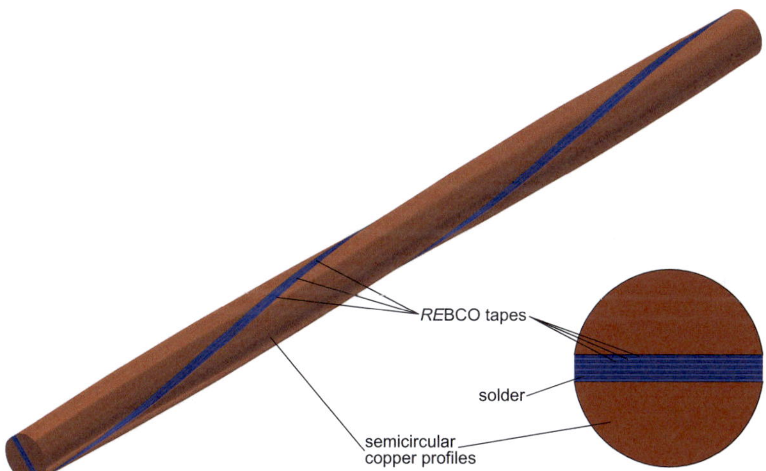

Figure 6.31: Schematic drawing of the first cabling stage of the Round Strands Composed of Coated Conductor Tapes (RSCCCT) cable concept. A stack of *RE*BCO tapes is soldered between two semicircular copper profiles. The number of tapes in the stack is low, resulting in a nearly circular conductor. Picture after [UWB13a].

The first stage is similar to the TSTC concept. The differences however, lie within the second stage of cabling. In the Round Strands Composed of Coated Conductor Tapes (RSCCCT) cable concept, the nearly circular conductors of the first stage of cabling (see **figure 6.31**) are

---

[k]EPFL-CRPP: http://crpp.epfl.ch/

stranded. Stranding of superconductors is a well established technology, used in many LTS cables. It completely transposes the strands.The second stage of cabling is shown schematically in **figure 6.32**.

Figure 6.32: Schematic drawing of the second stage of cabling of the Round Strands Composed of Coated Conductor Tapes (RSCCCT) cable concept. Several circular conductors of the cabling first stage are stranded.

However, as the *RE*BCO tapes are only partially transposed during the twisting of the first stage, the overall transposition of this concept remains partial if the self-field of the cable is taken into account [Bru13a]. Considering only magnetic background fields, the transposition of the second stage qualifies for complete overall transposition [Bru13b]. In fusion magnets, the magnetic self-field of each conductor can be neglected due to the high background fields.

At present, the information available on the mechanical properties is limited. Samples of the first cabling stage, consisting of 8, 3 mm wide, *RE*BCO tapes from SuperPower (SCS3050), exhibit a 2 % reversible degradation of the current carrying capabilities at bending radii of 0.5 m [UWB13b][UWB13a, p. 14]. A RSCCCT cable sample to be measured in-field is under construction by CRPP. A field and temperature dependent measurement is scheduled for 2014 within the European Fusion Development Agreement (EFDA) task.

6.6.2 Scaling to currents relevant for fusion magnets

By choosing an appropriate stranding pattern, the Round Strands Composed of Coated Conductor Tapes (RSCCCT) cable concepts can be scaled to very high currents. For the first stage of cabling, diverse sets of parameters are published [UWB13a, p. 14]:

- from 5 tapes of 3 mm width each, equaling 0.015 metre effective *RE*BCO width,

- to 20 tapes of 4 mm width each, equaling 0.08 m effective *RE*BCO width.

During first cabling stage of the Round Strands Composed of Coated Conductor Tapes cable concept the current carrying capabilities of the tapes are degraded about 1 % [UWB13a, p. 15]. The effective *RE*BCO width and the number of strands in second stage of cabling are calculated in **table 6.11** for fusion magnet conditions (68 kA at 12 T $\perp$) at 4.2 K, 22 K and 50 K cable temperature. The cross sectional area of the cable A_{RSCCCT} can be approximated from the number of subcables n_{subcable}, the number of tapes per subcable $n_{\mathrm{tapes\text{-}per\text{-}subcable}}$, the width w and thickness t of the *RE*BCO tapes using **equation 6.7**. A packing density v_{circle} of 90.69 % is assumed (hexagonal packing arrangement of circles).

$$A_{\mathrm{RSCCCT}} \approx \frac{\pi \cdot n_{\mathrm{subcable}}}{v_{\mathrm{circle}}} \cdot \left(w^2 + (0.5 \cdot t \cdot n_{\mathrm{tapes\text{-}per\text{-}subcable}})^2 \right) \qquad [6.7]$$

Table 6.11: Scaling of RSCCCT cables to fusion relevant currents (68 kA at 12 T $\perp$) at cable temperatures of 4.2 K, 22 K and 50 K.

cable temperature T	4.2 K	22 K	50 K
$J_{\mathrm{c}}(77\,\mathrm{K},\ \mathrm{self\text{-}field})$	250 A/(cm−width)	250 A/(cm−width)	250 A/(cm−width)
lift factor: $J_{\mathrm{c}}(T,\ 12\,\mathrm{T}\ \perp)/J_{\mathrm{c}}(77\,\mathrm{K},\ \mathrm{self\text{-}field})$	1.70	0.83	0.23
cabling degradation	1 %	1 %	1 %
$J_{\mathrm{e}}(T,\ 12\,\mathrm{T}\ \perp)$	420.8 A/(cm−width)	205.4 A/(cm−width)	56.9 A/(cm−width)
effective *REBCO* width	1.62 m	3.31 m	11.96 m
number of subcables$^{\alpha}$	21	42	150
number of subcables$^{\beta}$	108	221	797
cross sectional area $A_{\mathrm{RSCCCT}}{}^{\alpha}$	12.4 cm^2	24.7 cm^2	88.3 cm^2
cross sectional area $A_{\mathrm{RSCCCT}}{}^{\beta}$	33.9 cm^2	69.4 cm^2	250 cm^2

α 20 tapes per subcable, 4 mm wide
β 5 tapes per subcable, 3 mm wide

The twist pitch of the first stage of cable is, with 280 - 360 mm, below the stage 5 twist pitch of the ITER toroidal field coil conductors. The twist pitch of the second stage of cabling depends on the stranding patterns used. As a conductor in fusion magnets, the large copper fraction of Round Strands Composed of Coated Conductor Tapes cables is advantageous for electrical stabilization. However, the cross sectional area of this cable concept A_{RSCCCT} is very large, especially for low tape width w and low number of tapes per subcable $n_{\mathrm{tapes\text{-}per\text{-}subcable}}$.

6.6.3 Contacts

Due to similarities in stage one cabling of the TSTC concept and the Round Strands Composed of Coated Conductor Tapes cable concepts, the same methods of contacting can be used. Contacts for TSTCs are described in **subsection 6.5.6**.

6.7 Comparison of HTS cable concepts

The manufacturing, transposition, scaling to fusion relevant currents, electrical properties, mechanical properties and cooling of the major HTS cable concepts are compared in **table 6.12**.

Table 6.12: Comparison of major HTS cable concepts: Roebel Assembled Coated Conductor (RACC), Coated Conductor Rutherford Cables (CCRC), Conductor on Round-Core (CORC), Twisted Stacked-Tape Cables (TSTC) and Round Strands Composed of Coated Conductor Tapes (RSCCCT).

	RACC	CCRC	CORC	TSTC	RSCCCT
cable stages	1	2	1	2	2
manufacturing	complex	very complex	easy	medium	medium
cabling degradation	low (3 %)	high (13.3 %)$^\alpha$	zero (0 %)	low (3 %)	very low (1 %)
tape consumption	high (209 %)$^\beta$	very high (291 %)$^\gamma$	medium (141 %)$^\delta$	low (103 %)$^\varepsilon$	low (101 %)$^\varepsilon$
contacting method realized	"U-shape" or "stair"	"U-shape" or "stair"	cones	BSCCO - *RE*BCO connectors	BSCCO - *RE*BCO connectors
transposition	complete	complete	partial	self-field: partial ext. field: complete	self-field: partial ext. field: complete
transposition length	much larger than ITER	lower than ITER	much lower than ITER	comparable to ITER	comparable to ITER
scaling to fusion currents	not possible	possible, high cross section area necessary	possible, low cross section area necessary	possible, high cross section area necessary	possible, low cross section area necessary
J_e(12 T $\perp$)	high	low	high	low	medium
electrical stabilization	added: copper tapes	added: copper tapes or former	added: copper tapes in layers or central former	added: copper tapes or former	included: semicircular copper profiles
n-values	as single tapes	- no data -	as single tapes	low (8-14)	- no data -
mechanical properties	low: voids must be filled to prevent movement & distribute loads	- no data -	very good: high strength former	medium: voids must be filled to distribute loads	- no data -
strain effect	as single tapes	- no data -	negligible$^\delta$	- no data -	- no data -
cooling	very good: thermal transport along the tapes	very good: thermal transport along the tapes	challenging: layers thermally shield each other	good: low number of tapes per stack	good: low number of tapes per stack

α assuming the proposed layout, is mainly influenced by the out-of-plane bending radius $r_\text{out-plane}$.

β meander structures punched from 12 mm wide *RE*BCO tapes.

γ meander structures punched from 4 mm wide *RE*BCO tapes, using the proposed geometry.

δ assuming a winding angle of 45°.

ε in single TSTCs and RSCCCT cables.

Manufacturing and contacting

RACC cables are a Roebel assembly of meander structured *RE*BCO tapes. The punching of the tapes and their assembly are complex processes. The CCRC concept combines several RACC cables into one large cable. In this second cabling stage, the Rutherford method is used, further increasing the complexity of the process. CORC cables, are easy to manufacture by winding layers of *RE*BCO tapes around a central former in alternating directions. TSTCs and RSCCCT cables are manufactured in two stages. The processes are more complex than the manufacturing of CORC cables and less complex than of RACC cables and CCRCs.

During the manufacturing of some of the HTS cables the current carrying capabilities of the *RE*BCO tapes are reduced. With 13.3 % (calculated from [Kar12][BDF12a, p. 15]), these degradations are highest for the proposed CCRC geometry [SGG10]. For the other HTS cable concepts, the degradations are with 1 - 3 % significantly lower [GFK09][TCBM11, p. 4][UWB13a, p. 15]. In CORC cables, no degradation is observed during cabling.

Individually contacting the *RE*BCO tapes with homogeneous contact resistance is possible in all investigated HTS cable concepts. "Stair" or "double stair" shaped termination for RACC cables and CCRCs, cone shaped terminations in CORC cables and BSCCO - *RE*BCO connections in the TSTC and RSCCCT cable concept are promising contacting methods.

_RE_BCO tape consumption

The consumption of *RE*BCO tapes, depends on the cabling degradation, on punching and the orientation of the tapes. Due to the significant cabling degradation in the proposed geometry, the winding angle of $20°$ and the punching of the *RE*BCO tapes, the tape consumption is with 291 % highest in the CCRC concept. With 209 %, the tape consumption is also high in RACC cables, assuming 3 % cabling degradation and meander structures punched from 12 mm wide tapes. Due to zero cabling degradation and a winding angle of $45°$, there is medium *RE*BCO tape consumption of 141 % in CORC cables. With 103 % and 101 %, the tape consumption is lowest in single TSTCs and RSCCCT cables as in the concepts the tapes are neither punched nor wound around a former.

Transposition

The transposition of the *RE*BCO tapes is complete in all stages of cabling in RACC cables and CCRCs. In CORC cables, the tapes are arranged in layers; the transposition is only partial. However, no negative impact on the current distribution has been observed in short CORC cable samples. Current ramp rates up to $8.3\,\mathrm{kA\,s^{-1}}$ yield no increase of the cable voltage. In TSTCs and RSCCCT cables, the transposition of the superconductor tapes is also partial, considering magnetic self fields. For external fields, the transposition of the tapes is close to complete in both of these concepts.

Scaling to fusion relevant currents

All HTS cable concepts, excepts for RACC cables, can be scaled to the currents necessary for fusion magnets (e.g. 68 kA at 12 T background fields) at twist pitches of 0.42 m (stage five twist pitch of the ITER toroidal field coil conductor) and lower. RACC cables are subcables for the CCRC concept. The transposition lengths of CCRCs and CORC cables are significantly smaller than 0.42 m at fusion magnet boundary conditions while in TSTCs and RSCCCT cables they are comparable to the ITER twist pitch. Increasing the number of tapes marginally increases the cross sectional areas of RACC cables and CORC cables, yielding high engineering current densities J_e. In CCRCs, TSTCs and RSCCCT cables, higher number of subcables significantly increase the cross sectional area, reducing the engineering current density J_e.

Electrical properties

For safe operation, the amount of electrical stabilization in copper stabilized *RE*BCO tapes is insufficient. Because of this, additional electrical stabilization has to be included in all HTS cables. In RACC cables and CCRCs, meander structured copper tapes can be included and be assembled together with the *RE*BCO tapes. In CORC cables, either layers of copper tapes or formers with a high electrical conductivity are possible. On long lengths, layers of copper tapes are preferable as they improve the transfer of current from the tapes to the stabilization in the case of a quench. For TSTCs, electrical stabilization through additional copper tapes in the stacks has been proven to be effective. The copper fraction of RSCCCT cable is intrinsically very high due to the semicircular copper profiles on both sides of the superconductor tapes stacks. The electrical stabilization of RSCCCT cables is expected to be very high.

Superconducting transitions of RACC cables and CORC cables are sharp with n-values comparable to single *RE*BCO tapes. In the TSTC sample, lower n-values of $8 - 14$ are observed, due to irreversible degradation of the *RE*BCO tapes at high radial mechanical loads. At present, there is no n-value data available for in-field measurements of CCRCs and RSCCCT cables.

Mechanical properties

RACC cables are sensitive to mechanical loads if the *RE*BCO tapes are not fixated. The tapes can either be clamped or the voids can be filled, to prevent movement of the tapes and to distribute mechanical loads evenly. The impact of mechanical strains on the current carrying capabilities, the strain effect, is comparable to single tapes in RACC cables in the longitudinal direction [BBL12]. CORC cables, are mechanically strong with negligible strain effect due to the 45° winding angle (see **subsection A.1**). Repeated radial loading and unloading does not influence the cable performance. TSTCs also require mechanical stabilization of the superconductor tapes. A 12 % degradation of the current carrying capabilities has been observed in a TSTC cable in conduit conductor consisting of 40 *RE*BCO tapes during the first field increase. The degradation saturates and does not continue on subsequent load cycles. At present, there is no mechanical data for CCRCs and RSCCCT cables available.

Cooling

In RACC cables and CCRCs, all tapes repeatedly alternate their position between the center and the surface of the cable. The effective cooling contact area per tape is maximal at the surface. In the center, heat has to be transported through the tapes which is possible due to the high longitudinal thermal conductivity of *RE*BCO tapes. Because of this, the average cooling contact area of RACC cables and CCRCs is high, making them promising candidates for conduction cooling. In CORC cables, the layers of *RE*BCO tapes thermally shield each other due to the low perpendicular thermal conductivity of *RE*BCO tapes, strongly reducing the cooling contact areas in some layers. There are voids between the superconductor tapes which can be used as cooling channels. As these voids are also arranged in layers, there is little interchange of the cooling agent between the layers, rendering the cooling of CORC cables challenging. In TSTCs and RSCCCT cables, the number of *RE*BCO tapes per stack is low, homogenizing the thermal contact of all tapes. Due to the stranding of the second cabling stage of RSCCCT cables, forced flow cooling with low temperature margins is expected to be possible as in LTS fusion magnet conductors.

6.8 HTS winding packs for fusion magnets

To demonstrate the applicability and the benefits of HTS cables as conductors in fusion magnets, two exemplary winding packs utilizing HTS are calculated:

- a winding pack matching the boundary conditions of the inner legs of the toroidal field coils of ITER. This calculation is shown in **subsection 6.8.1**.

- an extrapolated winding pack for a fusion power plant. The parameters of power plant's coils are the current DEMO1 (2012) PROCESS output from the German DEMO Working Group [Kem12a, Kem12b]. It uses the same base for the materials and cross sectional areas of the coils as the previous one. However, the performance of the superconductor is extrapolated to higher fields. The extrapolations for LTS and for HTS are described in **subsection 6.8.2**.

Winding packs of the inner legs of the ITER toroidal field coils are chosen, as forces and available space are most critical in these areas. Only the winding pack itself is considered as the casing is independent of the magnet conductor technology used. An ITER toroidal field coil and the winding pack of the inner leg are shown schematically is **figure 6.33**. The HTS winding pack assumes a CORC cable as conductor. The CORC cable concept is selected as:

- there is no degradation of the current carrying capabilities during cabling; as shown in the in field tests, the current carrying capabilities of the *RE*BCO tapes are fully utilized (see **subsection 6.4.3.2**).

- it is resistant to mechanical loads. In the longitudinal and the radial direction, mechanical stresses have no influence on the current carrying capabilities in the stress range relevant for the conductor of fusion magnets.

- regardless of temperatures and magnetic background field, the superconducting transitions of CORC cables remain sharp.

- the scaling is easy. CORC cables can be scaled to very high numbers of tapes by adding more layers. The engineering current density J_e is high.

- there is no data available showing negative impact on the cable performance due to the partial transposition in CORC cables. Charging and discharging with high current ramp rates yield no inhomogeneous current distribution or reduction of the current carrying capabilities [LLG13, p. 23].

All HTS cable concepts, except for RACC cables, are viable as well.

6.8.1 HTS winding pack for ITER

For the calculation of the HTS winding pack, the operating parameters of the coils and the cross sectional areas of the different components are extracted from the ITER Design Description

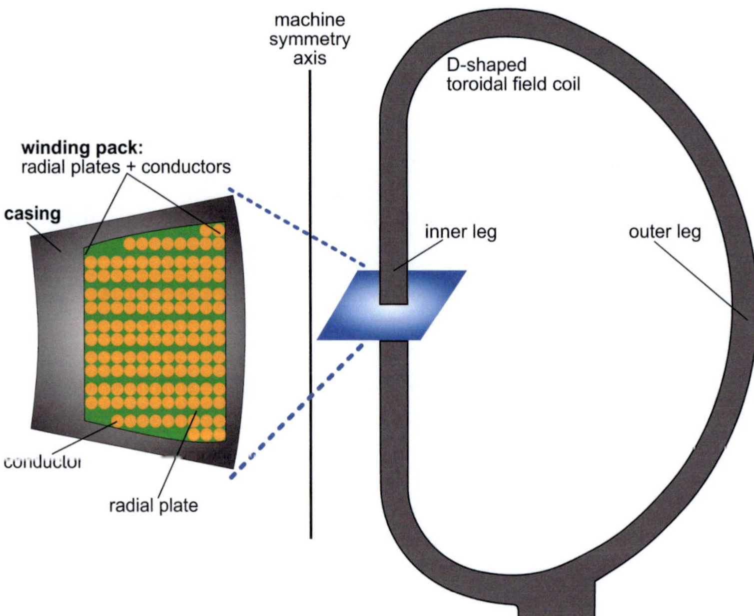

Figure 6.33: Schematic drawing of an ITER toroidal field coil and the winding pack of its inner leg.

Documents [ITE09b, ITE09e]. The operating parameters of the ITER toroidal field coils are given in **table 6.13**.

Table 6.13: Operating parameters of the toroidal field coils of ITER.

parameter	symbol	value
number of turns	$n_{\text{TF}}^{\text{ITER-LTS}}$	134
operating current	$I_{\text{op}}^{\text{ITER-LTS}}$	68 kA
safety margin, which is the quotient of operating current I_{op} and critical current I_{c} of the conductor: $f_{\text{safety}} = I_{\text{op}}/I_{\text{c}}$	$f_{\text{safety}}^{\text{ITER-LTS}}$	0.3
maximal magnetic field on the conductor	$B_{\text{max}}^{\text{ITER-LTS}}$	11.8 T

The main components of winding packs are the electrical stabilization, the superconductor, the structural materials and the voids for cooling. Their cross sectional areas are given in the following and summarized in **table 6.14**.

Total winding pack area

The HTS winding pack is compared against the total area of the LTS winding pack $A_{\text{total}}^{\text{ITER-LTS}}$ of $468\,000\,\text{mm}^2$.

Electrical stabilization

The total area of the electrical stabilization, including the copper matrix of the superconductor filaments, $A_{\text{Cu}}^{\text{ITER-LTS}}$ of $68\,100\,\text{mm}^2$ is kept constant in the HTS winding pack.

Structural materials

The total area reserved for structural materials, including the jacket of the superconductor cable and the radial plates, $A_{\text{structure}}^{\text{ITER-LTS}}$ of $267\,000\,\text{mm}^2$ is kept constant in the HTS winding pack.

Void area

The total void area $A_{\text{void}}^{\text{ITER-LTS}}$ of $56\,862.9\,\text{mm}^2$ is kept constant in the HTS winding pack. Due to its negligible cross sectional area, the electrical insulation is not considered.

Table 6.14: Cross sectional areas of the components of the LTS winding pack of the inner legs of the toroidal field coils of ITER.

component	symbol	cross sectional area	
total winding pack area	$A_{\text{total}}^{\text{ITER-LTS}}$	$468\,000\,\text{mm}^2$	reference for comparison
electrical stabilization	$A_{\text{Cu}}^{\text{ITER-LTS}}$	$68\,100\,\text{mm}^2$	kept constant in the HTS winding pack
structural materials	$A_{\text{structure}}^{\text{ITER-LTS}}$	$267\,000\,\text{mm}^2$	kept constant in the HTS winding pack
void area	$A_{\text{void}}^{\text{ITER-LTS}}$	$56\,862.9\,\text{mm}^2$	kept constant in the HTS winding pack

6.8.1.1 Calculation method

A CORC cable is scaled to the number of tapes necessary to carry the current in the windings $I_{\text{op}}^{\text{ITER-HTS}}$ of 68 kA using the same safety margin $(f_{\text{safety}}^{\text{ITER-HTS}} = f_{\text{safety}}^{\text{ITER-LTS}})$ of 0.3. The critical current of the CORC cable $I_{\text{c}}^{\text{ITER-HTS}}$ is calculated in **equation 6.8**.

$$I_{\text{c}}^{\text{ITER-HTS}} = \frac{I_{\text{op}}^{\text{ITER-HTS}}}{f_{\text{safety}}^{\text{ITER-HTS}}} = \frac{68 \text{ kA}}{0.3} = 226.67 \text{ kA} \qquad [6.8]$$

As there is no 11.8 T, 4.5 K data available for *REBCO* tapes, marginally higher fields on the conductor $B_{\text{max}}^{\text{ITER-HTS}}$ of 12 T are assumed while the temperature $T_{\text{op}}^{\text{ITER-HTS}}$ is reduced to 4.2 K. The field and temperature dependence and the current carrying capabilities of standard 2012 SuperPower *REBCO* tapes are used. The tapes are able to carry currents of 250 A/(cm−width) at 77 K in self field conditions (data from [Haz12, p. 8]). Their lift factor increases the critical current to 425 A/(cm−width) at 4.2 K and 12 T $\perp$ (see **figure 6.11**). The necessary effective superconductor width $w_{\text{eff}}^{\text{ITER-HTS}}$ can be calculated using **equation 6.9**.

$$w_{\text{eff}}^{\text{ITER-HTS}} = \frac{I_{\text{c}}^{\text{ITER-HTS}}}{I_{c}^{\text{HTS}}(4.2 \text{ K}, 12 \text{ T})} = \frac{226\frac{2}{3} \text{ kA}}{425 \text{ A/cm-width}} = 5.33 \text{ m-width} \qquad [6.9]$$

Self-field effects can be neglected at these background fields. This effective width corresponds to $n^{\text{ITER-HTS}}$ *REBCO* tapes with a width of $w = 4$ mm as calculated in **equation 6.10**.

$$n^{\text{ITER-HTS}} = \left\lfloor \frac{w_{\text{eff}}^{\text{ITER-HTS}}}{w} \right\rfloor = \left\lfloor \frac{5\frac{1}{3} \text{ m}}{4 \text{ mm}} \right\rfloor = 1333 \qquad [6.10]$$

Using the iterative method, described in the appendix in **section A.5**, the diameter d and the void area v of a CORC cable with 5.5 mm diameter former containing 1333 tapes are calculated. The tapes are arranged in 134 layers resulting in an outer diameter $d^{\text{ITER-HTS}}$ of 32.3 mm. The total area of this CORC cable $A_{\text{total-CORC}}^{\text{ITER-HTS}}$ is 819.4 mm^2. It contains a void area $A_{\text{void-CORC}}^{\text{ITER-HTS}}$ of 37.6 mm^2. Copper stabilized *REBCO* tapes from SuperPower (SCS4050) consist of a significant fraction of structural materials and of electrical stabilization as shown in **table 6.15**. In each tape, the thickness of the structural materials $t_{\text{structure}}$ is 0.05 mm and 0.044 mm for the electrical stabilization t_{Cu}. The 5.5 mm diameter former of the CORC cable is made of copper. Their cross sectional areas are calculated in **equation 6.11** and **equation 6.12**.

$$A_{\text{Cu-CORC}}^{\text{ITER-HTS}} = n^{\text{ITER-HTS}} \cdot w \cdot t_{\text{Cu}} + \left(\frac{d_1}{2}\right)^2 \cdot \pi \qquad [6.11]$$

$$= 1333 \cdot 4 \text{ mm} \cdot 0.044 \text{ mm} + \left(\frac{5.5 \text{ mm}}{2}\right)^2 \cdot \pi = 258.4 \text{ mm}^2$$

Table 6.15: Geometric parameters of the *REBCO* tapes used in the HTS winding pack calculation. Parameters correspond to 4 mm wide, copper stabilized SuperPower tapes (SCS4050).

component	symbol	value
width	w	4 mm
total tape thickness	t	0.1 mm
thickness of substrate (structural material)	$t_{structure}$	0.05 mm
thickness of electrical stabilization	t_{Cu}	0.044 mm

$$A_{\text{structure-CORC}}^{\text{ITER-HTS}} = n^{\text{ITER-HTS}} \cdot w \cdot t_{\text{structure}} \tag{6.12}$$

$$= 1333 \cdot 4 \text{ mm} \cdot 0.05 \text{ mm} = 266.6 \text{ mm}^2$$

At 1333 tapes, there is already significant cross sectional area of structural materials and electrical stabilization in the CORC cable. The calculated cross sectional areas of the CORC cable are per winding. For the whole winding pack they have to be multiplied with the number of turns $n_{\text{TF}}^{\text{ITER-HTS}}$ of 134. To reach the same total stabilization cross sectional area, total structure cross sectional area and total void cross sectional area as in the LTS winding pack, additional stabilization, structure and void area have to be added. Their corresponding cross sectional areas are calculated in **equation 6.13**, **equation 6.14** and **equation 6.15**.

$$A_{\text{Cu-add}}^{\text{ITER-HTS}} = A_{\text{Cu}}^{\text{ITER-LTS}} - n_{\text{TF}}^{\text{ITER-HTS}} \cdot A_{\text{Cu-CORC}}^{\text{ITER-HTS}} \tag{6.13}$$

$$= 68100.0 \text{ mm}^2 - 134 \cdot 258.4 \text{ mm}^2 = 33478.9 \text{ mm}^2$$

$$A_{\text{structure-add}}^{\text{ITER-HTS}} = A_{\text{structure}}^{\text{ITER-LTS}} - n_{\text{TF}}^{\text{ITER-HTS}} \cdot A_{\text{structure-CORC}}^{\text{ITER-HTS}} \tag{6.14}$$

$$= 267000.0 \text{ mm}^2 - 134 \cdot 266.6 \text{ mm}^2 = 231275.6 \text{ mm}^2$$

$$A_{\text{void-add}}^{\text{ITER-HTS}} = A_{\text{void}}^{\text{ITER-LTS}} - n_{\text{TF}}^{\text{ITER-HTS}} \cdot A_{\text{void-CORC}}^{\text{ITER-HTS}} \tag{6.15}$$

$$= 56862.9 \text{ mm}^2 - 134 \cdot 37.6 \text{ mm}^2 = 51822.5 \text{ mm}^2$$

Summation of the cross sectional areas of the CORC cables and the additional areas for the stabilization, the structure and the cooling (void), yields the total cross sectional area required by the HTS winding pack $A_{\text{total}}^{\text{ITER-HTS}}$ as shown in **equation 6.16**.

$$A_{\text{total}}^{\text{ITER-HTS}} = n_{\text{TF}}^{\text{ITER-HTS}} \cdot A_{\text{total-CORC}}^{\text{ITER-HTS}} + A_{\text{Cu-add}}^{\text{ITER-HTS}} + A_{\text{structure-add}}^{\text{ITER-HTS}} + A_{\text{void-add}}^{\text{ITER-HTS}} \tag{6.16}$$

$$= 134 \cdot 819.4 \text{ mm}^2 + 33478.9 \text{ mm}^2 + 231275.6 \text{ mm}^2 + 51822.5 \text{ mm}^2$$

$$= 426376.3 \text{ mm}^2$$

The winding pack obtained with this calculation method is an approximation to determine cross sectional areas. It is not designed as a model of an HTS fusion magnet coil. For an actual HTS coil, additional details have to be considered, e.g. several smaller CORC cables are much better suited to carry the current instead of a single one with that many layers.

6.8.1.2 Results

The total cross sectional area of the HTS winding pack for the inner legs of the toroidal field coils of ITER $A_{\text{total}}^{\text{ITER-HTS}}$ is $426\,376.3\ \text{mm}^2$. Compared with the total cross sectional area of the existing LTS winding pack $A_{\text{total}}^{\text{ITER-LTS}}$ of $468\,000\ \text{mm}^2$, this corresponds to a reduction of $8.9\,\%$. The amounts of electrical stabilization, structural materials, void area and the safety margin f_{safety} are kept constant. The LTS and HTS winding pack are compared schematically in **figure 6.34**.

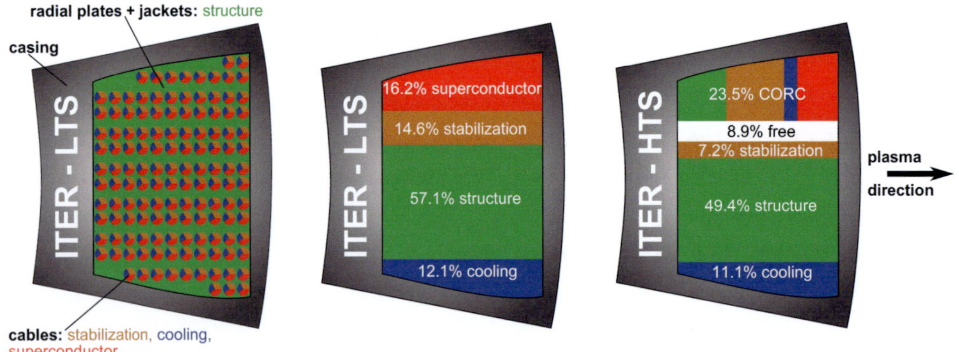

Figure 6.34: Schematic drawing of the winding pack of the inner legs of the ITER toroidal field coils (left). The cross sectional areas of the components (stabilization, structure, superconductor and void area) are analyzed (center) and compared with an HTS winding pack using the same boundary conditions (right).

The assumptions are conservative, e.g. the use of identical safety margins f_{safety} for LTS and HTS. In LTS cables, small increases of the conductor temperature strongly reduce their current carrying capabilities. The temperature dependence of REBCO is much lower near $4.2\ \text{K}$ (see **figure 3.12**), increasing the thermal stability of the HTS winding pack. This allows higher temperature margins and lower safety factors f_{safety}, further reducing the space required by HTS winding packs.

6.8.2 Winding packs for fusion power plants

In the following, two winding packs of a fusion power plant are extrapolated. The current DEMO1 (2012) PROCESS output [Kem12b] is used as the power plant's main parameters (see **table 3.2**). DEMO1's toroidal field coil parameters are an operating current per winding $I_{\mathrm{op}}^{\mathrm{DEMO1}}$ of 68 kA, a safety margin $f_{\mathrm{safety}}^{\mathrm{DEMO1}}$ of 0.3 and a maximal magnetic field on the conductor $B_{\mathrm{max}}^{\mathrm{DEMO1}}$ of 13.45 T as summarized in **table 6.16**. In a first step, the winding pack of the inner legs of the

Table 6.16: Operating parameters of the toroidal field coils of the fusion power plant concept DEMO1 (2012).

parameter	symbol	value
number of turns	$n_{\mathrm{TF}}^{\mathrm{DEMO1\text{-}LTS}}$	153
operating current	$I_{\mathrm{op}}^{\mathrm{DEMO1\text{-}LTS}}$	68 kA
safety margin	$f_{\mathrm{safety}}^{\mathrm{DEMO1\text{-}LTS}}$	0.3
maximal magnetic field on the conductor	$B_{\mathrm{max}}^{\mathrm{DEMO1\text{-}LTS}}$	13.45 T

toroidal field coils of ITER is adjusted to compensate for the increase in magnetic field on the conductor. The corresponding calculation is shown in **subsection 6.8.2.1**. In a second step, an HTS winding pack is calculated and compared with the LTS extrapolation in **subsection 6.8.2**.

6.8.2.1 Calculation method for LTS

In DEMO1 (2012) the maximal magnetic field on the conductor $B_{\mathrm{max}}^{\mathrm{DEMO1\text{-}LTS}}$ is increased by 14.0 % to 13.45 T. However, the current per winding $I_{\mathrm{op}}^{\mathrm{DEMO1\text{-}LTS}}$ is kept unchanged at 68 kA. Thus, the number of turns $n_{\mathrm{TF}}^{\mathrm{DEMO1\text{-}LTS}}$ has to be increased by the same factor as the field increases (14 %) to 153. The calculations are based on the cross sectional areas of the different components (stabilization, structural materials, superconductor and void area) of the ITER winding pack [ITE09b, ITE09e]. Electrical insulation is not considered.

Electrical stabilization

The electrical stabilization per winding is unchanged as the operating current is kept constant. However, a 14 % increase of the number of turns results in an increase of the total area of stabilization $A_{\mathrm{Cu}}^{\mathrm{DEMO1\text{-}LTS}}$ by the same factor to 77 756.0 mm^2.

Void area

The total void area $A_{\mathrm{void}}^{\mathrm{DEMO1\text{-}LTS}}$ is also increased proportional to the number of turns by 14 % to 64 925.6 mm^2.

Structural materials

A 14 % increase of magnetic field, increases the Lorentz forces proportionally and the necessary amount of structural materials per winding at constant current. However, the number of turns is increased by 14 %, too. Thus, the total cross sectional area of the structural materials $A_{\text{structure}}^{\text{DEMO1-LTS}}$ is enlarged by 30.0 % to $347\,486.7\,\text{mm}^2$.

Superconductor area

Increasing the magnetic field decreases the current carrying capabilities of the superconductor. For Nb_3Sn, the reduction is 39 % going from 12 T to 13.45 T background field as shown in the **appendix A.3** in **figure A.8**. The cross sectional area of the superconductor $A_{\text{superconductor}}^{\text{DEMO1-LTS}}$ is increased by that factor to $120\,686.2\,\text{mm}^2$.

Total winding pack area

The total area of the DEMO1 winding pack is the sum of the cross sectional areas of the stabilization, the void area, the area reserved for structural materials and the area required by the Nb_3Sn superconductors as shown in **equation 6.17**.

$$A_{\text{total}}^{\text{DEMO1-LTS}} = A_{\text{Cu}}^{\text{DEMO1-LTS}} + A_{\text{void}}^{\text{DEMO1-LTS}} + A_{\text{structure}}^{\text{DEMO1-LTS}} + A_{\text{superconductor}}^{\text{DEMO1-LTS}} \qquad [6.17]$$

$$= 610854.4 \ \text{mm}^2$$

This is an increase of the total winding pack area $A_{\text{total}}^{\text{DEMO1-LTS}}$ of 30.5 % compared with the ITER LTS winding pack $A_{\text{total}}^{\text{ITER-LTS}}$. The LTS winding pack of the fusion power plant is summarized and compared with ITER's winding pack in **table 6.17**.

Table 6.17: Cross sectional areas of the components of the LTS winding pack of the inner legs of the toroidal field coils of the fusion power plant concept DEMO1 (2012).

component	symbol	cross sectional area	compared with ITER	
stabilization	$A_{\text{Cu}}^{\text{DEMO1-LTS}}$	$77\,756.0\,\text{mm}^2$	114 %	kept constant in the HTS winding pack
structure	$A_{\text{structure}}^{\text{DEMO1-LTS}}$	$347\,486.7\,\text{mm}^2$	130 %	kept constant in the HTS winding pack
void area	$A_{\text{void}}^{\text{DEMO1-LTS}}$	$64\,925.6\,\text{mm}^2$	114 %	kept constant in the HTS winding pack
superconductor	$A_{\text{superconductor}}^{\text{DEMO1-LTS}}$	$120\,686.2\,\text{mm}^2$	139 %	-
total	$A_{\text{total}}^{\text{DEMO1-LTS}}$	$610\,854.4\,\text{mm}^2$	130.5 %	reference for comparison

6.8.2.2 Comparison with HTS

For the fusion power plant, the maximal magnetic field on the conductor $B_{\text{max}}^{\text{DEMO1-HST}}$ is increased by 14.0 % to 13.45 T. This reduces the current carrying capabilities of the *REBCO* tapes by

6.8 %, increasing the number of tapes required to 1440. This is used in combination with the cross sectional areas of the LTS DEMO1 extrapolation (see **subsection 6.8.2.1**), as input for the calculation method presented in **subsection 6.8.2**.

The total area of the HTS winding pack of the fusion power plant concept DEMO1 (2012) $A_{\text{total}}^{\text{DEMO1-HTS}}$ is estimated to 532594.7 mm^2. Compared with the total cross sectional area of the LTS extrapolation of a DEMO1 winding pack $A_{\text{total}}^{\text{DEMO1-LTS}}$, this corresponds to a reduction of 12.8 %. The LTS and HTS winding packs for DEMO1 (2012) are shown in a schematic drawing in **figure 6.35**.

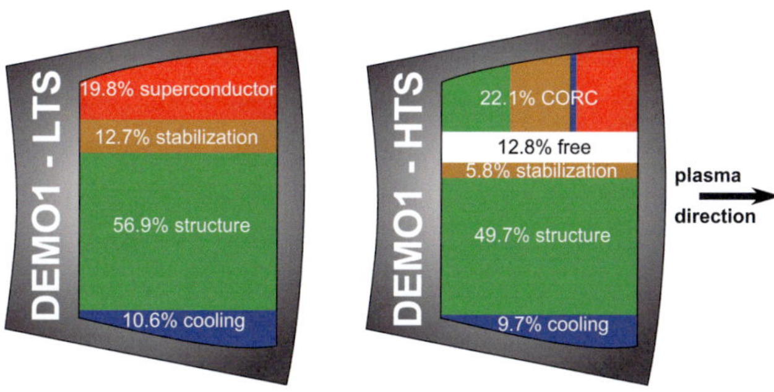

Figure 6.35: Schematic drawing of an extrapolated winding pack for the fusion power plant concept DEMO1 (2012) [Kem12b, Kem12a]. The ITER LTS winding pack is extrapolated to the boundary conditions of the fusion power plant (left). An HTS winding pack is calculated for these conditions (right).

The benefit of HTS on the size of the winding pack is increased by the boundary conditions of the fusion power plant DEMO1 (2012) compared with ITER. The impact of the field increase from 12 T to 13.45 T on the critical currents of the superconductors is with 6.8 % for *REBCO* compared with 39 % for Nb$_3$Sn, much less severe. Additionally, the mechanical properties of the superconductor are even more important at higher fields. To summarize, increasing the maximal magnetic fields always increases the benefits of HTS cables as conductor in the magnets.

6.8.3 Summary

Two HTS winding packs for fusion magnets are calculated and compared with existing LTS technology. The amounts of electrical stabilization, structural materials and voids for cooling are kept constant. The safety margin f_{safety} of the ITER toroidal field coils of 0.3 is used. The composition, current carrying capabilities and in-field performance of 2012 standard 4 mm copper stabilized SuperPower tapes (SCS4050) are assumed. The HTS winding packs utilizes CORC cables as a conductor, scaled to the necessary number of tapes.

The HTS winding pack for ITER assumes the operational parameters of the toroidal field coils (68 kA at 12 T $\perp$ background fields). The cross sectional areas for the different components (electrical stabilization, structural materials, superconductor and void areas) are extracted from the ITER Design Description Documents [ITE09b, ITE09e] for the inner legs of these coils. Electrical insulation is not yet considered. The results of these calculations are summarized and compared with the winding pack of ITER in **table 6.18**. Firstly, these calculations demonstrate that the requirements of large fusion magnets can be fulfilled with HTS assuming the long-length tape performance of 2012. At the same safety margin (0.3), the same amount of structural materials, electrical stabilization and void area, the total total cross sectional area of the winding pack is reduced by 8.9 % compared with the ITER LTS winding pack.

Secondly, LTS and HTS winding packs are extrapolated satisfying the boundary conditions of the fusion power plant concept DEMO1 (2012). In this concept the maximal magnetic field on the conductor is increased by 14 % to 13.45 T. The current per winding is kept constant, resulting in an 14 % increase in the number of turns to 153. The ITER LTS winding pack is extrapolated to these conditions. The cross sectional areas of the electrical stabilization, the structure and the void area are increased to compensate for the higher number of turns and Lorentz forces on the winding. The superconductor cross sectional area even has to be increased by 39 % due the strong reduction of the current carrying capabilities of Nb_3Sn at magnetic background fields of 13.45 T. Thus, the total cross sectional area of the LTS winding pack of the fusion power plant is 30.5 % larger compared with ITER conditions. An HTS winding pack satisfying these boundary conditions is calculated, too. In the HTS winding pack, the reduction of the current carrying capabilities due to the field increase is only 6.8 %. 1440 tapes SuperPower *REB*CO tapes (4 mm wide) are needed assuming the 2012 long-length tape performance. The LTS and the corresponding HTS winding packs for the fusion power plant DEMO1 are summarized and compared in **table 6.19**. Due to the negligible cross sectional area, electrical insulation is not yet considered.

Table 6.18: The cross sectional areas of the components of an HTS winding pack are compared with the existing LTS winding pack of ITER (inner legs of the toroidal field coils). Boundary conditions of ITER are assumed. Electrical insulation is not considered.

ITER	LTS	HTS			
		in the CORC	additional	total	compared with LTS
stabilization	$68\,100.0\,mm^2$	$34\,621.1\,mm^2$	$33\,478.9\,mm^2$	$68\,100.0\,mm^2$	100 %
structure	$267\,000.0\,mm^2$	$35\,724.4\,mm^2$	$231\,275.6\,mm^2$	$267\,000.0\,mm^2$	100 %
void area	$56\,862.9\,mm^2$	$5040.4\,mm^2$	$51\,822.5\,mm^2$	$56\,862.9\,mm^2$	100 %
superconductor	$76\,037.1\,mm^2$	1333 tapes	-	1333 tapes	-
total	$468\,000.0\,mm^2$	$109\,799.3\,mm^2$	$316\,577.0\,mm^2$	$426\,376.3\,mm^2$	91.1 %

With these assumptions, the total cross sectional area of the HTS winding pack is reduced by 12.8 % compared with the extrapolated LTS winding pack. At higher magnetic fields, the reduction of the cross sectional area will be even larger.

Table 6.19: The cross sectional areas of the components of an HTS winding pack are compared with an LTS winding pack, extrapolated to the boundary conditions of the fusion power plant concept DEMO1 (2012) [Kem12b, Kem12a]. Electrical insulation is not considered.

DEMO1 (2012)	LTS	HTS			
		in the CORC	additional	total	compared with LTS
stabilization	$77756.0\,\mathrm{mm^2}$	$42411.3\,\mathrm{mm^2}$	$35344.6\,\mathrm{mm^2}$	$77756.0\,\mathrm{mm^2}$	100 %
structure	$347486.7\,\mathrm{mm^2}$	$44064.0\,\mathrm{mm^2}$	$303422.7\,\mathrm{mm^2}$	$347486.7\,\mathrm{mm^2}$	100 %
void area	$64925.6\,\mathrm{mm^2}$	$5954.4\,\mathrm{mm^2}$	$58971.2\,\mathrm{mm^2}$	$64925.6\,\mathrm{mm^2}$	100 %
superconductor	$120686.2\,\mathrm{mm^2}$	1440 tapes	-	1440 tapes	-
total	$610854.4\,\mathrm{mm^2}$	$134856.2\,\mathrm{mm^2}$	$397738.5\,\mathrm{mm^2}$	$532594.7\,\mathrm{mm^2}$	87.2 %

7 Summary

It is mandatory in fusion reactors to use superconducting magnets to confine the plasma. At present, superconducting fusion magnets use the low temperature superconductors (LTS) niobium-titanium (NbTi) and niobium-tin (Nb_3Sn). In future fusion reactors, the second generation high temperature superconductors (HTS) rare-earth-barium-copper-oxide (*RE*BCO) tapes offer attractive benefits such as higher critical mechanical stress and strain limits (see left of **figure 2.10** and left of **figure 2.14**), lower mechanical strain effect (see right of **figure 2.10** and left of **figure 2.14**), higher engineering current density in very high magnetic background fields (see **figure 2.7**), lower impact of temperature increases on the critical current density (see **figure 3.12**) compared with Nb_3Sn.

The main objective of this work is the systematic investigation, assessment and comparison of HTS cable concepts in the frame of their applicability as conductor in future fusion magnets. For this, the cables are analyzed through experiments and simulations starting with their constituent materials. Thermal expansion and thermal conductivity of the materials of HTS cables are measured and compared, minimizing the thermal expansion mismatches to prevent thermal stresses while maximizing the thermal conductivity. By finite element method (FEM) simulations and by experiments, Roebel Assembled Coated Conductor (RACC) cables are optimized. Two RACC cables, one Conductor on Round Core (CORC) cables and one Twisted Stacked-Tape Cable (TSTC) are characterized at different temperatures and magnetic background fields using the temperature variable insert of the FBI (force F, field B, current I) test facility. The main results of this work are summarized as following.

The thermal expansions of structural materials such as austenitic stainless steels 316-LN, Hastelloy C-276 and Nitronic 40 match the thermal expansion of *RE*BCO tapes quite closely. The thermal expansions of insulation materials are much higher, especially for pure plastics, which are therefore unsuitable in HTS cables. The thermal expansions of composites, such as glass-fiber-reinforced-plastics (G10), are close to the expansion of *RE*BCO tapes and thus are preferred to pure plastics. Pure glues and resins exhibit strong thermal expansion mismatches, as well. These can be significantly reduced by mixing with a low expansion sediment, e.g. quartz powder. *RE*BCO tapes and cables can be impregnated with mixtures of one part Araldite epoxy resin and one part quartz powder without damaging the superconductor tapes as demonstrated on short samples and a 1.2 m long HTS cable consisting of 15 tapes. With a low thermal expansion mismatch and melting point in range of 117 - 126 °C, In50Sn50 solder is the preferred soft solder for HTS cables.

The thermal conductivities of *RE*BCO tapes and of the composite insulation materials are highly anisotropic. In *RE*BCO tapes, the thermal conductivity perpendicular to the tape surface is up to three orders of magnitude lower than the conductivity in parallel direction. This means layers of *RE*BCO tapes are thermally shielding each other, therefore reducing the effective cooling contact area. Because of this, the arrangement of the *RE*BCO tapes in an HTS cable determines the cooling. Fully transposed cable concepts, e.g. Roebel Assembled Coated Conductor (RACC) cables, exhibit significantly higher effective cooling contact areas compared with concepts in which the superconductor tapes are arranged in layers. In addition to the arrangement of the *RE*BCO tapes, the cooling contact is also determined by the filler. The thermal conductivities of glues and resins are two orders of magnitude below the thermal conductivity of In50Sn50 soft solder. Based on these material investigations, several HTS cable concepts are investigated in detail.

Firstly, in Roebel Assembled Coated Conductor (RACC) cables, the mechanical properties and the electrical performance are strongly determined by the shape of the meander structures, which is necessary for the cabling. In the meander structures, tensile loads result in stress concentrations near the inner corners, limiting the yield strength of RACC cables. The stress concentrations can be reduced while the minimal twist pitch per number of tapes is increased by adjusting the meander shape. Outer corners, inner radii of 6 mm, angles of 35° and 10 % wider crossing sections are ideal for high strength RACC cables, as they increase the longitudinal mechanical strength of the cable by at least 60 % and the minimal twist pitch by less than 25 %. Even with the optimized meander structure, RACC cables are sensitive to mechanical loads, especially perpendicular to the cable direction. Forces in the $10\,\mathrm{kN\,m^{-1}}$ range can already damage the *RE*BCO tapes and degrade the cable. To improve the mechanical properties of RACC cables, the tapes have to be fixed. This can be achieved by tightly clamping the cable into a support structure or by filling the voids between the tapes. A mixture of Araldite epoxy resin and quartz powder (1:1) was successfully used to impregnate RACC cables. However, the effective cooling contact area is drastically reduced due to the low thermal conductivity of this filler. The shape of the meander structures influences the electrical performance as well. Increasing meander structure's width by 10 % improves the total current of the RACC cable by 7 % while requiring the same amount of superconductor tape. The contacts of HTS cables are identified as a major source of inhomogeneities. Identical contact resistances are necessary in short samples or in direct current applications for all tapes to carry the same current. In RACC cables, stair-shaped contacts with equal contact areas and pressures during soldering are proposed, yielding contact resistances in the 80 - 120 $\mathrm{n\Omega\,cm^2}$ range and resulting in RACC cables with n-values comparable to single *RE*BCO tapes.

Secondly, the current carrying capabilities of a Conductor on Round Core (CORC) cable are measured at different magnetic background fields and temperatures. All superconducting transitions remain sharp with high n-values, regardless of the magnetic field or the cable temperature. The temperature dependence of the critical current of this cable is comparable to single *RE*BCO tapes. No degradation of the sample is observed after load cycling. Current ramp rates up to $8.3\,\mathrm{kA\,s^{-1}}$ cause neither increased voltages nor inhomogeneous current distribution in this 1.2 m long CORC sample even though the transposition of the tapes is only partial due to its layered layout.

Thirdly, a Twisted Stacked-Tape Cable (TSTC), filled with soft solder, is characterized at different fields and temperatures. A 11.9 % irreversible degradation of the current carrying capabilities is observed on the first field increase. The degradation saturates and does not continue on subsequent load cycles. Because of this degradation, the n-values of this samples are, at $8 - 14$, significantly lower compared with single tapes or the CORC cable sample. The two clamped BSCCO - *RE*BCO connectors of the TSTC sample work perfectly even at maximal field (12 T) or at maximal current (10 kA).

Five major HTS cable concepts are evaluated in the frame of their applicability as a fusion magnet conductor. All cable concepts, except RACC cables, can be scaled to the number of tapes necessary to carry 68 kA at 12 T background fields at twist pitches comparable to ITER (stage 5 twist pitch of the toroidal field coils: 0.42 m). RACC cables, however are sub-cables in the Coated Conductor Rutherford Cable (CCRC) concept, increasing the scalability to match the current of a fusion magnet conductor even at 50 K cable temperatures. The Conductor on Round Core (CORC) cable, exhibits the best mechanical properties and in-field performance. No degradation and sharp superconducting transitions with high n-values are observed regardless of field and temperature. However, due to the winding angle of 45°, 41.4 % additional *RE*BCO tape length are required in this cable concept. For a given number of tapes, the twist pitch of CORC cables is minimal compared to the other HTS cable concepts, because of the layered structure, the transposition is only partial. Furthermore, the manufacturing of CORC cables is easiest as no soldering or punching of *RE*BCO tapes is necessary.

Two HTS winding packs utilizing single CORC cables are calculated: the winding pack of the ITER TF coil's inner legs and the winding pack of a fusion power plant. With the same amount of electrical stabilization, structural materials, void area and the same safety factor (0.3), the total cross sectional area is reduced by 9.8 % compared with the ITER LTS winding pack. In the fusion power plant concept DEMO1 (2012), the total cross section area is reduced by 12.8 % by using HTS instead of LTS. This demonstrates that more compact fusion magnet coils can possible within the boundary conditions of ITER and in fusion power plants utilizing HTS cables.

This work was concentrated on material properties and the mechanical characterization and comparison of various HTS cables. The next step is to extend to the electro-magnetic behavior of HTS cables, e.g. to secure quench stability in magnets. The investigations will continue within a European Fusion Development Agreement (EFDA) task with field and temperature dependent measurements of a Round Strands Composed of Coated Conductor Tapes (RSCCCT) cable sample and with the design of HTS fusion magnets.

A Annex

In **section A.1**, the anisotropic strain effects of *RE*BCO single crystals and *RE*BCO tapes are described in detail. In **section A.2**, the influence of shear stress on the current carrying capabilities of BSCCO and *RE*BCO tapes is investigated. The magnetic field dependencies of Nb_3Sn and *RE*BCO are compared in **section A.3**. Scaling behavior and fitting parameters of Conductor on Round Core cables are given in **section A.4** and **section A.5**, respectively. In **section A.6** fitting parameters used for the Twisted Stacked-Tape Cable concept are shown.

A.1 Anisotropic strain effect of *RE*BCO tapes

The strain dependency of the current carrying capabilities of *RE*BCO tapes is anisotropic. There are directions in which mechanical strains strongly influence the critical current, while in other directions their effect is negligible.

A.1.1 Single *RE*BCO crystals

Such anisotropy is also present in single *RE*BCO crystals, as shown in **figure A.1**. If pressure is applied on the a-axis (100) of the crystal's unit cell, the critical temperature is reduced with increasing pressure. Pressure on the b-axis (010) of the unit cell increases the critical temperature T_c. Pressure on the c-axis (001) hardly influences the critical temperature.

As critical temperature and critical current are linked, changes in T_c also influence the current carrying capabilities in the same way.

A.1.2 *RE*BCO tapes

*RE*BCO tapes are not single crystals, they are aligned poly crystals instead. In the tapes, the c-axis (001) of all crystal's unit cells are always perpendicular to the tape surface. A-axis (100) and b-axis (010), are in the tape plane, either in tape direction or perpendicular to the tape direction. Orientation of a-axis and b-axis can change from grain to grain. If mechanical strain is applied along the tape, in some grains the strain acts on the a-axis. In other grains, the b-axis is strained instead. The critical temperature and critical current of the grains with the b-axis (010) is increased if compressive strain is applied in the tape direction. These grains are shown in

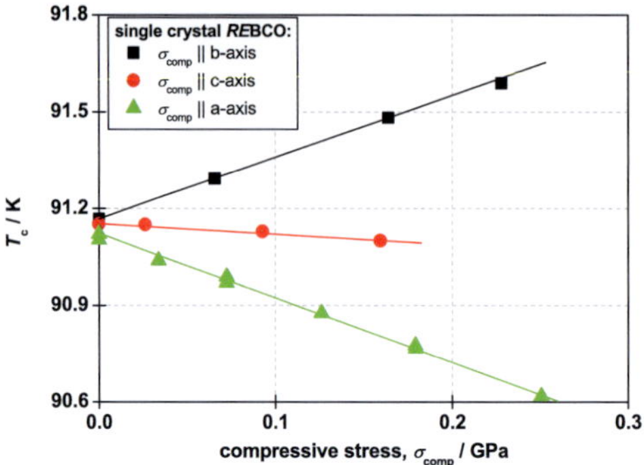

Figure A.1: Influence of compressive stress on the critical temperature of single *REBCO* crystals for different directions. Data from [WGF92].

green in the upper part of **figure A.2**. However, the critical temperature and the critical current in the grains with the a-axis (100) in tape direction is reduced at the same time. These grains limit the current carrying capabilities of the whole tape.

However, if mechanical strain is applied at an angle of 45° to the tape direction, it acts on the a-axis and b-axis of all grains in equal measure. Reductions (due to the a-axis strain) and increases (due to the b-axis strain) and of the critical temperature and critical current cancel each other out. In the grains with the a-axis in the tape direction and in the grains with the b-axis in the tape direction, the current carrying capabilities remain unchanged. This case is shown schematically in the bottom half of **figure A.2**.

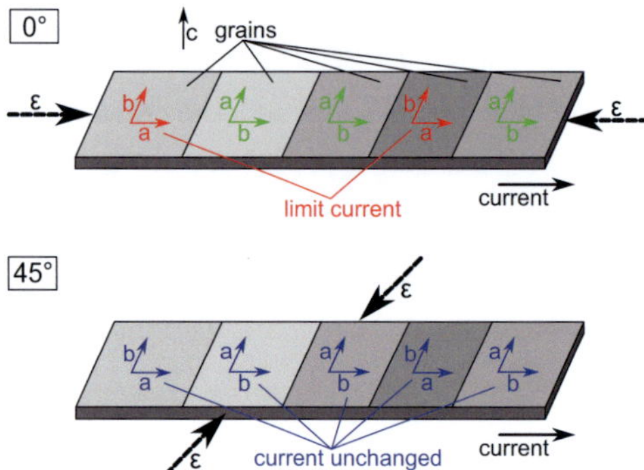

Figure A.2: Schematic drawing of the anisotropic strain effect of *REBCO* tapes. On the top: mechanical strain is parallel to the tapes (0°). Current is reduced in grains with a-axis (100) parallel to the tape. These grains (shown in red) limit the transport current. On the bottom: mechanical strain is applied in an angle of 45° to the tape. Strain acts in equal measure on a-axis (100) and b-axis (010). The transport current of the tape is not reduced. Picture after [LLG13].

This effect is observed in *REBCO* tapes in which small measurement sections are cut into the tapes at different angles [LAP11]. Mechanical pressure is applied to the tapes while the critical currents of the measurement sections are measured and compared. As shown in **figure A.3**, there are significant differences in the investigated orientations of 22.7°, 47.4° and 89.9°.

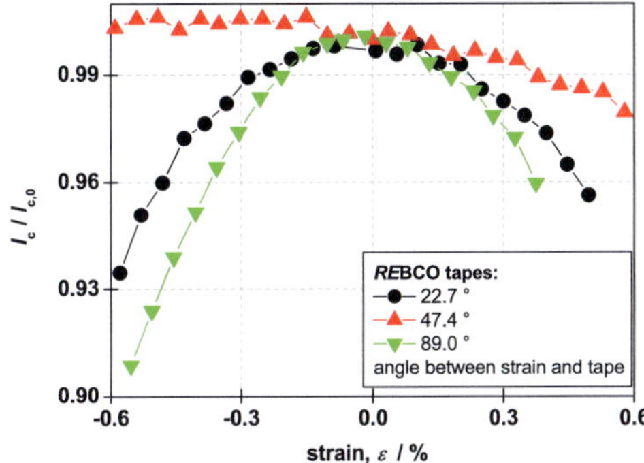

Figure A.3: Influence of mechanical strain on the current carrying capabilities of *RE*BCO tapes for different angles between the strain and the tape. Data from [LAP11].

At angles close to 45°, compressive and tensile strains hardly influence the current carrying capabilities. For smaller and larger angles the strain effect increases. At angles of 0° and 90° between the strain direction and the direction of the *RE*BCO tapes, the strain effect is maximal.

A.2 Influence of shear stress on the current carrying capabilities of HTS tapes

The current carrying of HTS tapes can be influenced by mechanical stresses. The magnitude of the degradation depends on the superconductor material and the direction and strength of the stresses. For compressive and tensile stresses, there is a lot of information available (see **subsection 2.3.3.2** and **subsection 2.3.1**). However, the impact of pure shear stress is unknown.

The current carrying capabilities of *RE*BCO from AMSC and BSCCO 2223 tapes from European High Temperature Superconductors (EHTS) are analyzed under applied shear stresses in the 0 - 50 MPa range. This investigation has already been published by the author in [BWG11].

A.2.1 Experimental setup

Shear stress is generated within 4 mm wide HTS tapes using a three point bending method. The tapes and the bending apparatus are submerged in a liquid nitrogen bath. The critical current of the tapes is measured at different shear stress levels.

The tape to be tested is glued between two stainless steel (1.4429) frames of 23 mm height, 300 mm length and the same width as the tape. A two component epoxides, LARIT 335 and

LARIT 340, is used for bonding. This resin is known to sustain very high shear stresses. Prior to gluing, the surface of the steel frames is roughened with glass bead blasting and chemically cleaned to maximize the bonding strength. Force is applied in a symmetrical 3-point arrangement with two contact points near the ends of the steel frame at 50 mm and 250 mm. In the center (at 150 mm) the opposing force is induced. This is a standardized geometry commonly used for the testing of glues, epoxides and insulation [AD]. It is known to exhibit areas with high uniform shear stress.

The two steel frames and the HTS tapes form a tight bending package. Applying the 3 point forces slightly bend that package. The top side of the bending package curves inwardly and it experiences compressive longitudinal strain towards its center. However, its bottom curves outwardly and exhibits a shift away from the center. The opposing directions of these shifts generate shear stress between both steel frames exactly where the superconducting tape is situated. The shear stress is transmitted through the epoxy resin to the HTS tape.

In this three point bending geometry, the absolute value of the shear stress σ_s in the tapes can be calculated using the relation given in the standard [AD], where $\alpha = 0.0462$ m is the total height of the bending package and is the $\beta = 0.004$ m the width.

$$|\sigma_s^{\text{standard}}| = 0.75 \cdot \frac{1}{\beta \cdot \alpha} \cdot F_{\text{ext}}$$

$$= 4058 \cdot F_{\text{ext}} \cdot \frac{1}{\text{m}^2}$$

A.2.2 Finite element method calculations

The bending setup is simulated in a three dimensional finite element model. Real dimensions and the mechanical properties of all constituent materials are used. The distribution of shear stress in the bending package is shown schematically in **figure A.4**.

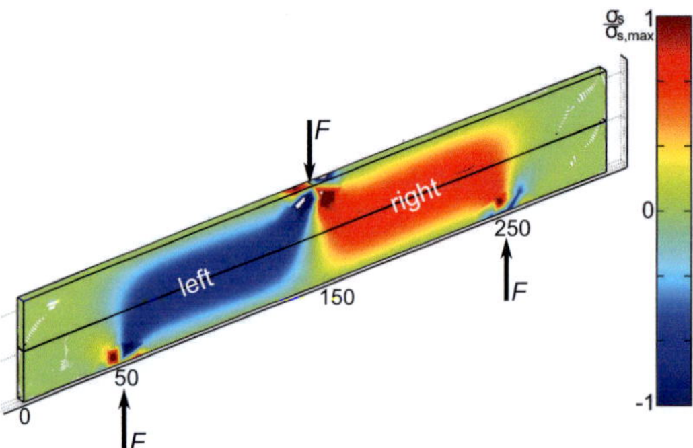

Figure A.4: Distribution of shear stress within a three point bending apparatus. FEM calculations reveal two zones with homogeneous shear stress: Negative shear stress (left) and positive shear stress (right).

Finite element simulations show two areas with homogeneous shear stress and low longitudinal stress. There is the left zone from 70 mm to 130 mm and the right zone from 170 mm to 230 mm. The directions of the shear stresses are reversed in the zones. In the left zone the shear stress is negative (colored in blue in **figure A.4**) and positive in the right zone (colored in red). Due to the symmetry of the setup, the zones are mirrored in the center (at 150 mm). The absolute values of the shear stresses are identical for both zones. These two homogeneous shear stress zones are the measurement sections of the apparatus. Small cavities of 1 mm edge length are machined into the steel frames, directly outside of these zones. Through the cavities, voltage taps are soldered to the HTS tape.

The shear stress distribution is simulated for different external forces, the averaged shear stress on the tape inside the measurement zones is calculated. The data is fitted with functions, determining the relations between external force F_{ext} and shear stress σ_{s}. Simulated data (points) and fitted data (lines) are shown in **figure A.5**.

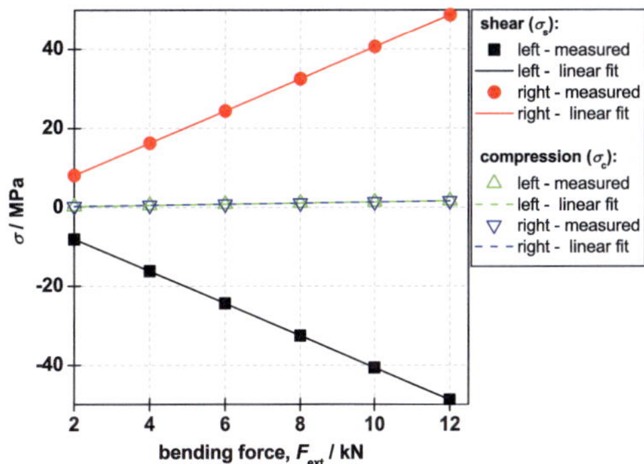

Figure A.5: Finite element calculation of the average shear stress and compressive stress in the measurement sections of the bending apparatus for different external forces. Simulated data (points) is fitted with linear functions (lines).

The absolute value of the shear stress can be determined using the relationship given in **equation A.1**.

$$|\sigma_{\text{s}}^{\text{simulation}}| = 4060 \cdot F_{\text{ext}} \cdot \frac{1}{\text{m}^2} \qquad [\text{A.1}]$$

The external force to shear stress relationships of the standard and of the simulations are in perfect agreement.

However, there is also a compressive stress component perpendicular to the tape surface. The relationship between the compressive stress σ_{c} and the external force F_{ext} is obtained from the finite element method calculations.

$$|\sigma_{\text{c}}| = 133 \cdot F_{\text{ext}} \cdot \frac{1}{\text{m}^2}$$

The compressive stress, however, is more than a factor of 30 smaller than the shear stress in both measurement sections. Thus, any influence of the compressive stress on the current carrying capabilities of the HTS tape can be neglected.

A.2.3 Results

Two different HTS tape types are characterized. Silver lined BSCCO tapes manufactured by EHTS and stainless steel laminated *RE*BCO tapes from AMSC. With a critical longitudinal tensile stress of 152 MPa, the specified mechanical properties of the BSCCO 2223 tapes are much lower than that specified for the *RE*BCO tapes. The *RE*BCO tapes are specified by the manufacturer with a critical longitudinal tensile stress of 360 MPa.

The current carrying capabilities of these tapes are measured at different external bending forces. **Equation A.1** is used to determined the shear stress in the tapes. The superconducting transitions at different shear stresses are shown in **figure A.6** for the BSCCO tapes (on the left) and for the *RE*BCO tapes (on the right).

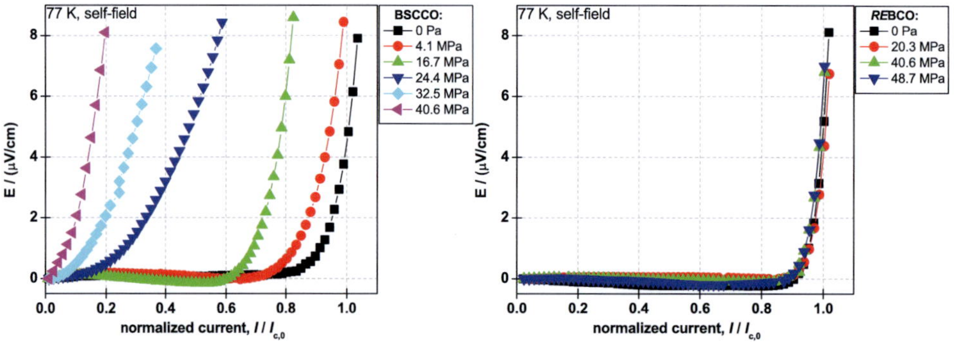

Figure A.6: Superconducting transitions of BSCCO tapes from EHTS (left) and *RE*BCO tapes from AMSC (right) at different shear stresses.

The current carrying capabilities of the BSCCO 2223 tapes are strongly reduced with increasing shear stress. Additionally, the steepness of the superconducting transitions is reduced. The critical currents of the *RE*BCO tapes are not influenced in the measured shear stress range 0 - 48.7 MPa. Higher shear stresses cannot be reached with this experimental setup as the epoxy resin starts to break loose from the HTS tapes at about 50 MPa. Using a $1\,\mu\text{V cm}^{-1}$ criteria, the normalized current carrying capabilities are shown in dependence of the applied shear stress in **figure A.7** for both sample types.

The reduction of the current carrying capabilities of the BSCCO 2223 tapes is linear with the applied shear stress.

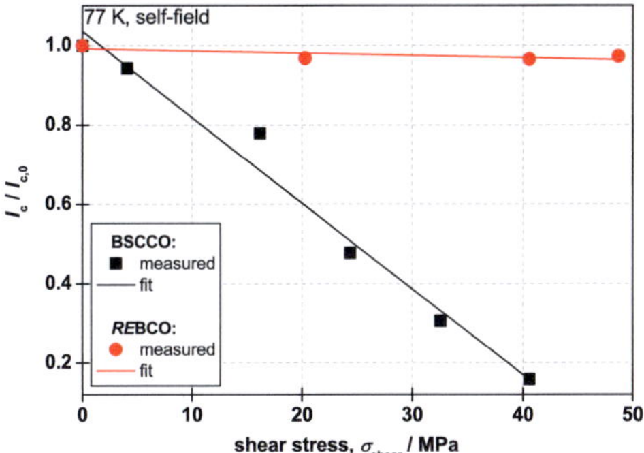

Figure A.7: Dependence of critical current on applied shear stress for BSCCO 2223 tapes from EHTS and REBCO tapes from AMSC. Measurements are done at 77 K in self-field conditions. The critical currents are determined using a $1\,\mu V\,cm^{-1}$ criteria.

A.3 Normalized critical currents of Nb$_3$Sn and REBCO at different magnetic fields

Magnetic fields influence the current carrying capabilities of superconductors. The magnitude of that influence depends not only on the temperature, but also on the superconductor material. In **figure A.8**, Nb$_3$Sn (internal tin, data from [Lee11]) and REBCO tapes from SuperPower (perpendicular field orientation, data from [Haz12]) are compared at 4.2 K. The current carrying capabilities are normalized, using background fields of 12 T as point of reference for both materials.

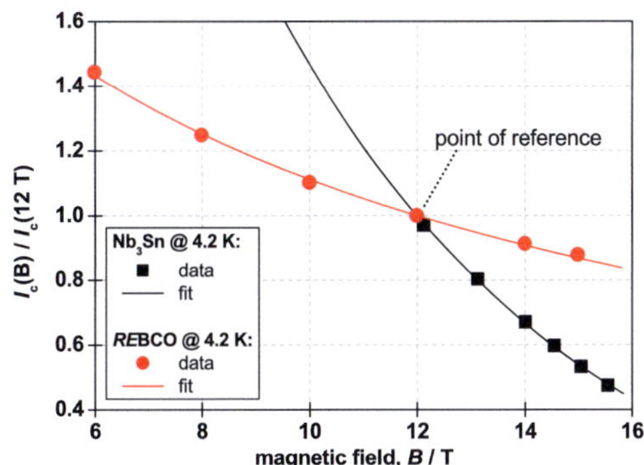

Figure A.8: Normalized current carrying capabilities of Nb$_3$Sn and REBCO tapes (perpendicular field) at 4.2 K and different magnetic background fields. A background field of 12 T is used as reference. Measured data (points) and fitted data (lines) are shown. REBCO data from [Haz12] and Nb$_3$Sn data from [Lee11].

Measured data is fitted with a function of exponential decay for Nb$_3$Sn and a reciprocal function for *REBCO*. Measured and fitted data are in good agreement for both superconductor materials. The normalized current carrying of Nb_3Sn at 4.2 K can be described in the magnetic field range of 11 - 15 T, using:

$$I_c^{Nb_3Sn}(4.2\ K, B)/I_c^{Nb_3Sn}(4.2\ K, 12\ T) = 9.125 \cdot \exp\left(\frac{-B}{5.7285}\right) - 0.126$$

The field dependence of *REBCO* tapes at 4.2 K can be described with:

$$I_c^{REBCO}(4.2\ K, B)/I_c^{REBCO}(4.2\ K, 12\ T) = \frac{0.586}{(0.231 + 0.0295 \cdot B)}$$

A.4 Fitting parameters of the field and temperature dependent tests of CORC cables

Table A.1 lists the parameters of reciprocal functions ($I_c = 1/(\alpha + \beta \cdot B)$) describing the correlation between critical current and magnetic background field of a CORC cable consisting of 15 SuperPower tapes. Measurements are performed in the FBI test facility using the temperature variable insert.

Table A.1: Parameters of reciprocal functions ($I_c = 1/(\alpha + \beta \cdot B)$) describing the correlation between critical current and magnetic background field of a CORC cable consisting of 15 SuperPower tapes.

$T_{surface}$	4.2 K	12.5 K	15.2 K	18.1 K	21.4 K
α	$8.80 \cdot 10^{-5}$	$9.86 \cdot 10^{-5}$	$1.03 \cdot 10^{-4}$	$1.08 \cdot 10^{-4}$	$1.13 \cdot 10^{-4}$
β	$2.71 \cdot 10^{-5}$	$2.92 \cdot 10^{-5}$	$3.18 \cdot 10^{-5}$	$3.62 \cdot 10^{-5}$	$4.37 \cdot 10^{-5}$

$T_{surface}$	25.0 K	29.0 K	33.5 K	38.3 K
α	$1.19 \cdot 10^{-4}$	$1.24 \cdot 10^{-4}$	$1.32 \cdot 10^{-4}$	$1.58 \cdot 10^{-4}$
β	$5.54 \cdot 10^{-5}$	$7.35 \cdot 10^{-5}$	$9.43 \cdot 10^{-5}$	$1.14 \cdot 10^{-4}$

A.5 Scaling behavior of CORC cables

The scaling of CORC cables to high number of tapes has to be calculated iteratively for each layer.

The circumference c_i is calculated from the diameter d_i of the current layer i, starting with the diameter of the former (commonly $d_1 = 5.5\,$mm).

$$c_i = \left(\frac{d_i}{2}\right)^2 \cdot \pi$$

The number of tapes n_i fitting in the circumference of the layer c_i depends on the width of the tapes w and the winding angle α.

$$n_i = \left\lfloor \frac{c_i}{w \cdot \cos(\alpha)} \right\rfloor$$

The void area v_i of the current layer i is defined as the fraction of the overall area of the layer a_i not used by the *REBCO* tapes.

$$a_i = \pi \cdot \left(\frac{(d_i + 2 \cdot t)^2 - d_i^2}{4} \right)$$

$$v_i = (c_i - n_i \cdot w \cdot \cos(\alpha)) \cdot t$$

t is the thickness of the *REBCO* tapes. At the next layer $(i+1)$, the diameter of the cable increases.

$$d_{i+1} = d_i + 2 \cdot t$$

The calculation has to be repeated for each layer. For the total number of *REBCO* tapes in the cable $n_{total}(k)$ at a specific layer k and the total void area $v_{total}(k)$ a summation over all inner layers $(1 \leq i \leq k)$ is necessary. The twist pitch of a CORC cable l_k with a specific number of layers k is derived from the winding angle α, the width of the tapes w and the number of tapes in the outer layer n_k.

$$n_{total}(k) = \sum_{i=1}^{k} n_i$$

$$v_{total}(k) = \sum_{i=1}^{k} v_i$$

The twist pitch of a CORC cable l_k with a specific number of layers k is derived from the winding angle α, the width of the tapes w and the number of tapes in the outer layer n_k.

$$l_k = n_k \cdot w \cdot sin(\alpha)$$

The total number of tapes and the total void area are calculated with this iterative method for a CORC cable consisting SuperPower tapes (width w of 4 mm and thickness t of 0.1 mm) wound at an angle of 45° around a former of 5.5 cm diameter. The data up to layer 215 is shown in 5 layer steps in **table A.2**. The layer numbers used in the scaling investigation of CORC cables to fusion relevant currents (**subsection 6.4.6** and **section 6.8**) are shown as well.

Table A.2: Scaling of CORC cables to fusion relevant currents assuming a former of 5.5 mm diameter, a winding angle of 45° and 4 mm wide SuperPower tapes.

layers	diameter / mm	tapes	void area / mm^2	layers	diameter / mm	tapes	void area / mm^2
1	5.7	3	0.03	110	27.5	948	30.69
5	6.5	15	0.80	115	28.5	1023	32.10
10	7.5	31	2.61	120	29.5	1101	33.37
15	8.5	51	3.72	125	30.5	1181	35.09
20	9.5	73	5.27	130	31.5	1265	36.12
25	10.5	98	6.69	134	32.3	1333	37.62
30	11.5	126	7.99	135	32.5	1350	38.15
35	12.5	156	9.72	140	33.5	1440	38.92
40	13.5	190	10.76	145	34.5	1531	40.70
42	13.9	204	11.39	150	35.5	1626	41.78
45	14.5	225	12.81	155	36.5	1723	43.30
50	15.5	265	13.59	160	37.5	1823	44.70
55	16.5	306	15.39	165	38.5	1926	45.97
60	17.5	351	16.49	170	39.5	2031	47.68
65	18.5	398	18.02	175	40.5	2140	48.70
70	19.5	448	19.44	180	41.5	2250	50.73
75	20.5	501	20.72	185	42.5	2365	51.49
80	21.5	556	22.44	190	43.5	2481	53.26
85	22.5	615	23.47	195	44.4	2601	54.34
90	23.5	675	25.51	200	45.5	2723	55.86
95	24.5	740	26.29	205	46.5	2848	57.25
100	25.5	806	28.07	210	47.5	2976	58.52
101	25.7	820	28.17	211	47.7	3002	58.73
105	26.5	876	29.16	215	48.5	3106	60.22

A.6 Fitting parameters of the field and temperature dependent tests of TSTCs

Table A.3 lists the parameters of reciprocal functions ($I_c = 1/(\alpha + \beta \cdot B)$) describing the correlation between critical current and magnetic background field of a CORC cable consisting of 15 SuperPower tapes. Measurements are performed in the FBI test facility using the temperature variable insert.

Table A.3: Parameters of reciprocal functions ($I_c = 1/(\alpha + \beta \cdot B)$) describing the correlation between critical current and magnetic background field of a CORC cable consisting of 15 SuperPower tapes.

T_{surface}	4.2 K	21.6 K	23.8 K	26.5 K	29.8 K	33.5 K	29.8 K
α	$5.37 \cdot 10^{-5}$	$4.81 \cdot 10^{-5}$	$4.65 \cdot 10^{-5}$	$4.01 \cdot 10^{-5}$	$4.09 \cdot 10^{-5}$	$4.24 \cdot 10^{-5}$	$4.14 \cdot 10^{-5}$
β	$1.24 \cdot 10^{-5}$	$1.65 \cdot 10^{-5}$	$1.83 \cdot 10^{-5}$	$2.09 \cdot 10^{-5}$	$2.25 \cdot 10^{-5}$	$2.44 \cdot 10^{-5}$	$2.71 \cdot 10^{-5}$

B Designations and abbreviations

1.4429	a stainless steel type
1.4471	a stainless steel type
1D	one dimensional
2D	two dimensional
316-LN	a stainless steel type
3D	three dimensional
A15	a unit cell structure in crystals
a	(100) direction of the crystal's unit cell
ABAD	alternating beam assisted deposition
AC	alternating current
Ag	silver
AgMg	silver-magnesium
AMSC	American Superconductor
Al	aluminum
Al-99.5	aluminum with a level of purity of 99.5 %
a.p.	advanced pinning
avg	average
b	(010) direction of the crystal's unit cell
Ba	barium
BCS	Bardeen, Cooper and Schrieffer
BSCCO	Bismuth-strontium-calcium-copper-oxide
c	(010) direction of the crystal's unit cell
c	critical
c	compressive
Ca	calcium
cc	coated conductor
CCRC	Coated Conductor Rutherford Cable
Ce	cerium
CERN	Conseil Européen pour la Recherche Nucléaire
CICC	cable-in-conduit conductors

CORC	Conductor on Round Core
Cr	chromium
CRPP	Centre de Recherches en Physique des Plasmas
CryoMaK	Cryogenic Material Tests Karlsruhe
CS	Central Solenoid
CSD	chemical solution deposition
Cu	copper
DC	direct current
DEMO	demonstration reactor
DT-670	a temperature sensor type
EDDC	evaporation using drum in dual chamber
EFDA	European Fusion Development Agreement
eff	effective
EPFL	École Polytechnique Fédérale de Lausanne
EUROFER	a reduced-activation-ferritic-martensitic steel
ext	external
FBI	force - field - current
Fe	iron
FEM	finite element method
FWD	first wall dose
G10	a common glass-fiber-reinforced-plastic type
G10 normal	direction of G10, perpendicular to the glass mattes
G10 warp	direction of G10, in plane with the glass mattes
GCS	General Cable Superconductor
H_2	deuterium
H_3	tritium
He	helium
HF	Helical Field
HTS	high temperature superconductor
IBAD	ion beam assisted depositioning
in	inner
In	indium
In50Sn50	solder consisting of 50 % indium and 50 % tin
int	internal
IRL	Industrial Research Limited
ISD	Inclined substrate deposition
ITEP	Institute for Technical Physics

ITER	International Experimental Thermonuclear Reactor
JET	Joint European Torus
KIT	Karlsruhe Institute of Technology
LHD	Large Helical Device
LN_2	liquid nitrogen
LTS	low temperature superconductor
max	maximal
MgB_2	Magnesium-diboride
min	minimal
MIT	Massachusetts Institute of Technology
Mo	molybdenum
MOD	metal organic deposition
MOVCD	metal organic chemical vapor deposition
MRI	magnetic resonance imaging
n	neutron
Nb	niobium
NbTi	niobium-titanium
Nb_3Al	niobium-aluminum
Nb_3Sn	niobium-tin
Ni	nickel
NMR	nuclear-magnetic-resonance
norm	normalized
op	operating point
op	operating
OST	Oxford Superconducting Technologies
out	outer
ov	overall
PA	polyamide
PAIR	pre-annealing and intermediate rolling
PEEK	polyether etherketone
PEI	polyetherimide
PC	polycarbonate
PES	polyethersulfone
PF	Poloidal Field
PID	proportional - integral - differential
PLD	pulsed laser depositioning
PP	polypropylene

PPE	polyphenylene ether
PPSU	polyphenylsulfone
PPMS	Physical Property Measurement System
PSU	polysulfone
PUR-SS	polyurethanes reinforced with stainless steel wires
quartz powder	fine grained powder of amorphous silica dioxide
RABITS	rolling assisted bi-axial textured substrate
RACC	Roebel Assembled Coated Conductor
RCE-DR	reactive co-evaporation by deposition and reaction
rel	relative
*RE*BCO	rare-earth-barium-copper-oxide
rel. width	geometry parameter relative width w_{rel} of the meander structure of RACC cables
req	requirement
RRR	residual resistivity ratio
RSCCCT	Round Strands Composed of Coated Conductor Tapes
RT	room temperature
s	shear
SC	superconductor
SCS12050	copper stabilized *RE*BCO tapes from SuperPower, 12 mm wide
SCS3050	copper stabilized *RE*BCO tapes from SuperPower, 3 mm wide
SCS4050	copper stabilized *RE*BCO tapes from SuperPower, 4 mm wide
s.f.	self-field
SF12050	stabilization free *RE*BCO tapes from SuperPower, 12 mm wide
Sn	tin
TC	thermal conductivity
TE	thermal expansion
TF	Toroidal Field
tot	total
TSTC	Twisted Stacked-Tape Cable
TTO	Thermal Transport Option
USA	United States of America
VAC	Vacuumschmelze
W	tungsten
Y	yttrium
YBCO	yttrium-barium-copper-oxide
Zr	zirconium

C Index of symbols

A	area, (m^2)
$\vec{A}_\mathrm{B}$	vector potential
A_CCRC	cross sectional area of CCRCs, (m^2)
A_CORC	cross sectional area of CORC cables, (m^2)
A_Cu	area of the stabilization, (m^2)
a_i	area of the layer i of CORC cables, (m^2)
$a_\mathrm{n}(T_\mathrm{n})$	calibration factor of an extensometers at the temperature T_n, (m V)
a_p	minor plasma radius, (m)
A_RACC	cross sectional area of RACC cables, (m^2)
A_RSCCCT	cross sectional area of RSCCCT cables, (m^2)
a_RT	calibration factor of an extensometers at the room temperature, (m V)
$A_\mathrm{structure}$	area of the structural materials, (m^2)
$A_\mathrm{superconductor}$	area of the superconductor, (m^2)
A_total	total area, (m^2)
A_TSTC	cross sectional area of TSTCs, (m^2)
A_void	void area, (m^2)
α	a fitting parameter
α	winding angle of HTS cables, (°)
α_B	a material constant
$\alpha_\mathrm{in\text{-}plane}$	in plane winding angle of CCRCs, (°)
B	magnetic field, (T)
B_0	a material constant, (T)
B_{0_ab}	a fitting parameter of the field dependence in ab-direction, (T)
B_{0_c}	a fitting parameter of the field dependence in c-direction, (T)
B_{0_i}	a fitting parameter of the internal field dependence, (T)
B_c	critical magnetic field, (T)
B_c1	lower critical magnetic field, (T)
B_c2	upper critical magnetic field, (T)
B_ext	external magnetic field, (T)
B_i	magnetic field in a conductor, (T)

B_{int}	internal magnetic field, (T)
B_{loc}	local magnetic field, (T)
B_{max}	maximal magnetic field on the conductor, (T)
B_{t}	magnetic field in the plasma, (T)
β	a fitting parameter
c	speed of light in vacuum, ($299\,792\,458\,\text{m s}^{-1}$)
c_{i}	circumference of CORC cables with i layers, (m)
C_{wires}	correction polynom of the wires in thermal conductivity measurements using the PPMS, (W K)
d	diameter, (m)
d_1	diameter of the former of CORC cables, (m)
d_{i}	diameter of a CORC cable with i layers, (m)
δ_{ab}	parameter describing the magnetic field in ab-direction, (°)
δ_{c}	parameter describing the magnetic field in c-direction, (°)
δ_{i}	parameter describing the internal magnetic field, (°)
Δ_{int}	distance between the plasma and the TF coils, commonly referred to as internal distance, (m)
Δj	current reserve, (A m^{-2})
ΔL	change in length of a material, (m)
ΔL_{TE}	change in length of a material due to its thermal expansion, (m)
ΔT	temperature reserve, (K)
Δx	measurement distance, (m)
e	electron charge, ($1.6022 \times 10^{-19}\,\text{C}$)
E	energy, (J)
E	electric field , (V m^{-1})
E_{c}	critical electric field , (V m^{-1})
ε	strain, (%)
ε	emissivity, (%)
$\varepsilon_{\text{TE}}\,(\text{RT} \rightarrow 4.2\,\text{K})$	thermal expansion of a material between room temperature and 4.2 K, (%)
$\varepsilon_{\text{TE}}\,(\text{RT} \rightarrow 77\,\text{K})$	thermal expansion of a material between room temperature and 77 K, (%)
$\varepsilon_{\text{long}}$	longitudinal strain, (%)
ε_{TE}	thermal expansion coefficient, (%)
$\varepsilon_{\text{trans}}$	transversal strain, (%)
F	force, (N)
f_{ab}	fitting function for magnetic fields in ab-direction
f_{c}	fitting function for magnetic fields in c-direction
F_{ext}	external force, (N)

f_i	fitting function for the internal magnetic field
F_L	Lorentz force, (N)
$F_{L,max}$	maximal Lorentz force on the conductor, (N m^{-1})
F_p	pinning force, (N)
f_{safety}	safety factor, $\frac{I_{op}}{I_c}$
γ	a fitting parameter
γ	angle between the direction of the magnetic field and the *REBCO* tape's surface normal vector
H_{ext}	absolute value of the external $\vec{H}$ field, (T)
I	current, (A)
I_c	critical current, (A)
$I_{c,0}$	critical current prior to impregnation, (A)
I_{coil}	overall current in the coil current, commonly referred to as the coil current, (A)
I_{ind}	induced current, (A)
$I_{ind,1}$	a specific induced current, (A)
$I_{ind,2}$	another specific induced current, (A)
I_{op}	operating current, (A)
j	current density, (A m^{-2})
J_{0_p}	a fitting parameter for the critical current density, (A m^{-2})
J_{0_i}	a fitting parameter for the critical current density, (A m^{-2})
j_c	critical current density, (A m^{-2})
$j_{c,norm}$	normalized critical current density
$J_{C_{ab}}$	function describing the critical current density in an applied field in *ab*-direction, (A m^{-2})
J_{C_c}	function describing the critical current density in an applied field in *c*-direction, (A m^{-2})
J_{C_i}	function describing the critical current density depending on the internal magnetic field, (A m^{-2})
J_e	engineering current density, (A m^{-2})
$j_{in}^\star$	incoming heat flux, (W m^{-2})
$j_{out}^\star$	outgoing heat flux, (W m^{-2})
J_T	transport current density, (A m^{-2})
k	number of layers of a specific CORC cable
κ	a fitting factor
$\kappa_n(T_n)$	correction factor of the calibration of an extensometer for the temperature T_n
l	length, (m)

l	piece length of *REBCO* tapes, (m)
L_0	length at room temperature, (m)
l_k	minimal twist pitch of CORC cables with k layers, (m)
u_{ab}	a fitting parameter of the angular field dependence
u_c	a fitting parameter of the angular field dependence
u_i	a fitting parameter of the angular field dependence
λ	thermal conductivity, (W K^{-1} m^{-1})
λ_e	electronic part of the thermal conductivity, (W K^{-1} m^{-1})
λ_p	phononic part of the thermal conductivity, (W K^{-1} m^{-1})
m	mass, (kg)
m	parameter of smoothing
μ_0	permeability, ($4\pi \times 10^{-7}$ N A^{-2})
n	the n-values is a measure for the steepness of the transition form superconducting state to normal conduction
n	number of tapes
n_i	maximal number of tapes in CORC cables with i layers.
n_k	maximal number of tapes in CORC cables with k layers.
$n_{subcable}$	number of subcables
$n_{tapes\text{-}per\text{-}subcable}$	number of tapes per subcable
n_{TF}	number of turns in the TF coils
n_{total}	total number of tapes in a specific CORC cable
v_{circle}	packing density of circles, (90.69 %)
O	surface, (m^2)
Q	heating power, (W)
Q_{rad}	radiation losses, (W)
r	radius, (m)
r_{in}	inner radius in the meander structure of RACC cables, (m)
r_{out}	outer radius in the meander structure of RACC cables, (m)
$r_{out\text{-}plane}$	out of plane bending radius of CCRCs, (m)
R_p	major radius, (m)
s	postition along *REBCO* tapes, (m)
s_{req}	space required for each crossing of meander structures in RACC cables, (m)
σ	stress, (Pa)
σ	Stefan-Boltzmann constant, (5.670400×10^{-8} J s^{-1} m^{-2} K^{-4})
σ_c	compressive stress, (Pa)
$\sigma_{\varepsilon_{TE}}$	standard deviation of thermal expansion measurements
σ_{long}	longitudinal stress, (Pa)

σ_{s}	shear stress, (Pa)
σ_{trans}	transversal stress, (Pa)
t	thickness, (m)
T	temperature, (K)
T_{average}	average temperature, (K)
T_{c}	critical temperature, (K)
$T_{\mathrm{c,0}}$	a definition of the critical temperature T_{c}, (K)
$T_{\mathrm{c,50}}$	a definition of the critical temperature T_{c}, (K)
$T_{\mathrm{c,onset}}$	a definition of the critical temperature T_{c}, (K)
t_{Cu}	total thickness of the electrical stabilization in REBCO tapes, (m)
T_{cs}	current sharing temperature, (K)
T_{n}	a specific temperature, (K)
T_{op}	operating temperature, (K)
T_{RT}	room temperature, (approximately 295 - 300 K)
$t_{\mathrm{structure}}$	thickness of the substrate in REBCO tapes, (m)
T_{surface}	surface temperature, (K)
θ	angle between the straight sections and the crossing sections in the meander structure of RACC cables, (°)
θ_{ext}	angle of the the external $\vec{H}$ field, (°)
θ_{loc}	angle of the local magnetic field B_{loc}, (°)
v	a fitting parameter of the angular field dependence
v	void area, (m^2)
v_{i}	void area of CORC cables with i layers, (m^2)
v_{total}	total void area of a specific CORC cable, (m^2)
w	width, (m)
w_{crossing}	width of the crossing sections in the meander structure of RACC cables, (m)
w_{eff}	effective superconductor width, (m)
w_{rel}	geometry parameter relative width of the meander structure of RACC cables
w_{straight}	width of the straight sections in the meander structure of RACC cables, (m)
w_{subcable}	width of subcables, (m)
x	distance, (m)
x	oxygen concentration
ξ	coherence length, (m)
ξ_{ab}	coherence length in ab-directions, (m)
ξ_{c}	coherence length in c-direction, (m)

D Bibliography

[ABB84]   ALLADIO, F ; BARBATO, E ; BARDOTTI, G ; BARTIROMO, R ; BRACCO, G ; BUCETI, G ; BURATTI, P ; CRISANTI, F ; ANGELIS, R D. ; MARC, F D. ; PRETIS, M D. ; FRIGIONE, D ; GASPAROTTO, M ; GIANNELLA, R ; GROLLI, M ; LENOCI, M ; MANCUSO, S ; MAZZITELLI, G ; PERICOLI, V ; PIERONI, L ; PODDA, S ; RIGHETTI, G B. ; ROMANELLI, F ; SANTINI, F ; SEGRE, S E. ; TUCCILLO, A A. ; ZANZA, V: Heating experiments on the FT Tokamak in the lower hybrid frequency range. In: *Plasma Physics and Controlled Fusion* 26 (1984), 1A, S. 157

[ABB06]   ABB. *1910-1919 Roebel bars and turbines. In:* Corporate Chronicle. 2006

[ABC81]   AYMAR, R ; BON MARIDON, G ; CLAUDET, G ; DISDIER, F ; DUTHIL, R ; GENEVEY, P ; HAMELIN, J ; LIBEYRE, P ; MAYAUX, G ; MEURIS, C ; PARAIN, J ; SEYFERT, P ; TOROSSIAN, A ; TURCK, B: Tore Supra: Status Report concerning the superconducting magnet after the qualifying development program. In: *IEEE Transactions on Magnetics* 17 (1981), 5, S. 1911–1914

[Abr57]   ABRIKOSOV, A A.: On the Magnetic Properties of Superconductors of the Second Group. In: *Journal of experimental and theoretical physics of the Academy of Sciences of the USSR* 5 (1957), 6, S. 1174–1182

[AD]   ASTM-D2344. *Standard Test Method for Short-Beam Strength of Polymer Matrix Composite Materials and Their Laminates*

[Alb04]   ALBRECHT, J: Low-Angle Grain Boundaries of YBCO in External Magnetic Fields. In: JOHANSEN, T H. (Hrsg.) ; SHANTSEV, D V. (Hrsg.): *Magneto-Optical Imaging.* Springer Netherlands, 2004. – ISBN 978–1–4020–1998–2, Kapitel 2, S. 159–165

[All]   ALLOY WIRE INTERNATIONAL. *Hastelloy C-276 Specification. In:* http://www.alloywire.com/german/hastelloy_alloy_c-276.html

[Ame]   AMERICAN SUPERCONDUCTOR. *Brass Laminated Amperium®Wire. In:* http://www.amsc.com/library/BRSAMP_DS_A4_0912.pdf

[AMS12]   AMSC. *Guidelines for Termination of Amperium Wire. In:* $http://www.amsc.$ $com/library/AMPTERM\_AN\_A4\_0112.pdf$. 2012

[AP08]   ADAMVALLI, M ; PARAMESWARAN, V: Dynamic strength of adhesive single lap joints at high temperature. In: *International Journal of Adhesion and Adhesives* 28 (2008), September, 6, S. 321–327. – ISSN 01437496

[AUI11]   AOKI, T ; UEDA, H ; ISHIYAMA, A ; MIYAHARA, N ; KASHIMA, N ; NAGAYA, S: Effect of Neutron Irradiation on High-Temperature Superconductors. In: *Applied Superconductivity, IEEE Transactions on* 21 (2011), 3, S. 3200–3202

[Bag12]   BAGRETS, N: Measurement uncertainty of thermal conductivity measurements. In: *personal communication* (2012)

[Bal02]   BALLARINO, A: HTS current leads for the LHC magnet powering system. In: *Physica C: Superconductivity* 372-376 (2002), August, S. 1413–1418. – ISSN 09214534

[Bar09]   BARTH, C: *Aufbau eines Messstandes zur Bestimmung von Schottky-Barrierenhöhen nicht-idealer Dioden*, University of Karlsruhe, Diploma Thesis, 2009

[Bay]   BAYER, C: Soldering methods of coated conductor tapes. In: *to be published*

[Bay12]   BAYER, C: High temperature superconducting current leads: An essential component of efficient superconducting applications. In: *KIT Fusion PhD workshop.* Bad Herrenalb, 2012

[BBBW13]   BAGRETS, N ; BARTH, C ; BAYER, C M. ; WEISS, K-P: Thermal and thermomechanical properties of soft solders for superconducting applications. In: *MEM11 Workshop.* Aix en Provence, 2013

[BBL12]   BADCOCK, R ; BUMBY, C ; LONG, N: HTS Roebel cable: Practical current injection and tensile strength. In: *Applied Superconductivity Conference.* Portland : Industrial Research Limited, 2012

[BBW13]   BARTH, C ; BAGRETS, N , WEISS, K-P , BAYER, C M. , BAST, T. Degradation free epoxy impregnation of REBCO coils and cables. In: *Superconductor Science and Technology* 26 (2013), 5

[BC11]   BOCCACCINI, L V. ; CISMONDI, F: Technology Parameters of in-vessel components that are to be considered in System Code Analyses / KIT. KIT CN, Karlsruhe, 2011. – Forschungsbericht. – 0–13 S

[BCS57]   BARDEEN, J ; COOPER, L N. ; SCHRIEFFER, J R.: Theory of Superconductivity. In: *Physical Review* 108 (1957), 5, S. 1175–1204

[BDF12a]   BARTH, C ; DROTZIGER, S ; FIETZ, W H. ; GOLDACKER, W ; HELLER, R ; WEISS, K-P: Final-Report on Construction and test of DEMO relevant cables (WP12-DAS-01-T06) / EFDA Work Program 2012 Design Assessment Studies: Superconducting Magnets (WP12-DAS01-MAG). Karslruhe Institute of Technology, CN, Eggenstein-Leopoldshafen, 2012. – Forschungsbericht

[BDF12b]   BARTH, C ; DROTZIGER, S ; FIETZ, W H. ; GOLDACKER, W ; HELLER, R ; WEISS, K-P: WP11-DAS-HTS - 02-01 KIT Assessment of HTS material, scalable cabling concepts / Karlsruhe Institute of Technology (KIT). Eggenstein-Leopoldshafen : Karlsruhe Institute of Technology, 2012. – Forschungsbericht

[Bea62]   BEAN, C P.: Magnetization of Hard Superconductors. In: *Phys. Rev. Lett.* 8 (1962), 6, S. 250–253

[Ber]   BERLINCOURT, T G. *Niobium-Titanium: Workhorse Supermagnet Material. In: http://magnet.fsu.edu/$\sim$lee/superconductor-history_ files/Nb-Ti/Niobium-Titanium--Workhorse_Supermagnet_ Material-2011.pdf*

[Bet39]   BETHE, H A.: Energy Production in Stars. In: *Phys. Rev. Lett.* 55 (1939), 5, S. 434–456

[BGJW12]   BAGRETS, N ; GOLDACKER, W ; JUNG, A ; WEISS, K-P: Thermal properties of REBCO copper stabilized superconducting tapes. In: *Applied Superconductivity Conference.* Portland, 2012

[BGS13]   BAGRETS, N ; GOLDACKER, W ; SCHLACHTER, S I. ; BARTH, C ; WEISS, K-P: Thermal properties of materials for Coated Conductor Rutherford Cables (CCRC). In: *submitted for publication* (2013)

[BH63]   BERLINCOURT, T G. ; HAKE, R R.: Superconductivity at High Magnetic Fields. In: *Phys. Rev. Lett.* 131 (1963), 1, S. 140–157

[BM86]   BEDNORZ, J G. ; MÜLLER, K A.: Possible High T_c Superconductivity in the Ba-La-Cu-O System. In: *Zeitschrift für Physik B, Condensed Matter* 64 (1986), 2, S. 189–193

[Boc11]   BOCCACCINI, L V.: DEMO technology code (PPPT). In: *3rd Meeting of the German DEMO working group.* Jülich, 2011

[Bru06]   BRUZZONE, P: 30 Years of Conductors for Fusion : A Summary and Perspectives. In: *Applied Superconductivity, IEEE Transactions on* 16 (2006), 2, S. 839–844

[Bru12]   BRUCKER.   *Bruker HTS YBCO Coated Conductor Data Sheet. In: http://www.bruker-est.com/fileadmin/be_user/Advanced_ Supercon/EHTS/products/YBCO/YBCO_Data_Sheet.pdf.* 2012

[Bru13a]   BRUZZONE, P: Transposition of twisted superconductor stacks at magnetic self-fields. In: *personal communication* (2013)

[Bru13b]   BRUZZONE, P: Transposition of twisted superconductor stacks in magnetic background fields. In: *personal communication* (2013)

[BTK08]   BARZI, E ; TURRIONI, D ; KIKUCHI, A ; LAMM, M ; RUSY, A ; YAMADA, R ; ZLOBIN, A V. ; BALACHANDRAN, U ; AMM, K ; EVANS, D ; GREGORY, E ; LEE, P J. ; OSOFSKY, M ; PAMIDI, S ; PARK, C ; WU, J ; SUMPTION, M: Bscco-2212 Wire and Cable Studies. In: *AIP Conference Proceedings* 986 (2008), 3, S. 431–438. – ISSN 0094243X

[Buc94]   BUCKEL, W: *Supraleitung: Grundlagen und Anwendungen*. 5th. Weinheim : VCH, 1994. – ISBN 978–3527290871

[BWG11]   BARTH, C ; WEISS, K-P ; GOLDACKER, W: Influence of Shear Stress on Current Carrying Capabilities of High Temperature Superconductor Tapes. In: *Applied Superconductivity, IEEE Transactions on* 21 (2011), 3, S. 3098–3101

[BWGS11]   BARTH, C ; WEISS, K-P ; GOLDACKER, W ; SCHLACHTER, S I.: Electromagnetic measurements of coated conductor cables at different temperatures. In: *MEM11 Workshop*. Okinawa, 2011

[BWVS12]   BARTH, C ; WEISS, K-P ; VOJENČIAK, M ; SCHLACHTER, S I.: Electromechanical analysis of Roebel cables with different geometries. In: *Superconductor Science and Technology* 25 (2012), Februar, 2, S. 025007. – ISSN 0953–2048

[BWWW12]   BAGRETS, N ; WEISS, E ; WESTENFELDER, S ; WEISS, K-P: Cryogenic Test Facility CryoMaK. In: *IEEE Transactions on Applied Superconductivity* 22 (2012), Juni, 3, S. 9501204–9501204. – ISSN 1051–8223

[Cav98]   CAVE, J R.: Critical Temperature. In: *Handbook of applied superconductivity*. Bristol : Inst. of Physics Publ., 1998. – ISBN 0750303778, S. 291–294

[CEA07]   CEA: Superconducting magnets for fusion. In: *CLEFS CEA - No. 56 - Magnets and magnetic materials* (2007)

[CETX07]   CHEGGOUR, N ; EKIN, J W. ; THIEME, C L H. ; XIE, Y Y.: Effect of fatigue under transverse compressive stress on slit Y-Ba-Cu-O coated conductors. In: *Applied Superconductivity, IEEE Transactions on* 17 (2007), 2, S. 3063–3066

[Che93]   CHERNOPLEKOV, N A.: The system and test results for the Tokamak T-15 magnet. In: *Fusion Engineering and Design* 20 (1993), S. 55–63

[CKW11]   CHAPMAN, I T. ; KEMP, R ; WARD, D J.: Analysis of high β regimes for DEMO. In: *Fusion Engineering and Design* 86 (2011), März, 2-3, S. 141–150. – ISSN 09203796

[CPCW08]   CHEN, W ; PAN, Y ; CHEN, Z ; WEI, J: The design and the manufacturing process of the superconducting toroidal field magnet system for EAST device. In: *Fusion Engineering and Design* 83 (2008), Januar, 1, S. 45–49. – ISSN 09203796

[Cul12]   CULHAM CENTER FOR FUSION ENERGY. *Introduction to fusion. In: http: //www. ccfe. ac. uk/ introduction. aspx.* 2012

[Dev11]   DEVRED, A. *Update on the ITER Conductor Production.* 2011

[DFH97]   DARWESCHSAD, M ; FRIESINGER, G ; HELLER, R ; IRMISCH, M ; KATHOL, H ; KOMAREK, P ; MAUER, W ; NÖTHER, G ; SCHLEINKOFER, G ; SCHMIDT, C ; SIHLER, C ; SÜSSER, M ; ULBRICHT, A ; WÜCHNER, F: Development and test of the poloidal field prototype coil POLO at the Forschungszentrum Karlsruhe. In: *Fusion Engineering and Design* 36 (1997), S. 227–250

[DH08]   DUCHATEAU, J L. ; HERTOUT, P: Which superconducting material for the toroidal field system of the fusion DEMO reactor? In: *Journal of Physics: Conference Series* 97 (2008), Februar, 1, S. 012038. – ISSN 1742–6596

[DHH79]   DEIS, D ; HENNING, C ; HINKLE, R ; KOPYTOFF, V ; MACDONALD, J: Fabrication of the MFTF magnet windings. In: *IEEE Transactions on Magnetics* 15 (1979), Januar, 1, S. 534–537. – ISSN 0018–9464

[Die06]   DIETRICH, R: 100m YBCO tape produced by EHTS / Brucker EST. 2006. – Forschungsbericht

[DKE06] DKE DEUTSCHE KOMMISSION ELEKTROTECHNIK ELEKTRONIK INFORMATIONSTECHNIK IM DIN UND VDE. *IEC 61788-3 Ed.2: Superconductivity - Part 3: Critical current measurement - DC critical current of Ag- and/or Ag alloy-sheathed Bi-2212 and Bi-2223 oxide superconductors.* 2006

[DKE07a] DKE DEUTSCHE KOMMISSION ELEKTROTECHNIK ELEKTRONIK INFORMATIONSTECHNIK IM DIN UND VDE. *IEC 61788-1 Supraleitfähigkeit - Teil 1: Messen des kritischen Stromes - Kritischer Strom (Gleichstrom) von Nb-Ti Verbundsupraleitern.* 2007

[DKE07b] DKE DEUTSCHE KOMMISSION ELEKTROTECHNIK ELEKTRONIK INFORMATIONSTECHNIK IM DIN UND VDE. *IEC 61788-2 Supraleitfähigkeit - Teil 2: Messen des kritischen Stromes - Kritischer Strom (Gleichstrom) von Nb_3Sn Verbundsupraleitern.* 2007

[DPG11] DPG DEUTSCHE PHYSIKALISCHE GESELLSCHAFT: Helium: rar und kostbar. In: *PHYSIKonkret Nr. 10* (2011), September, S. 1–2

[Dre95] DRESNER, L: *Stability of superconductors.* New York : Plenum Press, 1995. – ISBN 0–306–45030–5

[Dru00] DRUDE, P: The electron theory of metals. In: *Ann. Phys.* 1 (1900), S. 566–613

[Duc] DUCHATEAU, J L.: Dimensioning Magnetic Systems for Tokamaks: From ITER to DEMO. In: *Matefu Spring School*, Association Euratom-CEA

[Dup] DUPONT. *Summary of the properties of Kapton® polyimide films. In:* `http://www2.dupont.com/Kapton/en_US/assets/downloads/pdf/summaryofprop.pdf`

[ECB71] EFFERSON, K R. ; COFFEY, D L. ; BROWN, R L. ; DUNLAP, J L. ; GAUSTER, W F. ; LUTON, J N. ; SIMPKINS, J E.: The IMP Superconducting Coil System. In: *Nuclear Science, IEEE Transactions on* 18 (1971), 4, S. 265–272

[EFD] EFDA. *Fusion conditions. In:* `http://www.efda.org/fusion/how-fusion-works/fusion-conditions/`

[Ein05a] EINSTEIN, A: Ist die Trägheit eines Körpers von seinem Energiegehalt abhängig? In: *Annalen der Physik* 323 (1905), 13, S. 639–643

[Ein05b] EINSTEIN, A: Zur Elektrodynamik bewegter Körper. In: *Annalen der Physik* 322 (1905), 10, S. 891–921

[Eis07]   EISTERER, M: Magnetic properties and critical currents of MgB 2. In: *Superconductor Science and Technology* 20 (2007), Dezember, 12, S. R47–R73. – ISSN 0953–2048

[EL63]   EL BINDARI, A. ; LITVAK, M. M.: Upper Critical Field of Nb-Zr and Nb-Ti Alloys. In: *Journal of Applied Physics* 34 (1963), 9, S. 2913. – ISSN 00218979

[EL93]   EDELMAN, H S. ; LARBALESTIER, D C.: Resistie transitions and the origin of the n-value in superconductors with a Gaussian critical-current distribution. In: *J. Appl. Phys* 74 (1993), 5, S. 3312–3315

[Emea]   EMERSON & CUMING. *Stycast® 2850 FT - Thermally Conductive Epoxy Encapsulant. In: http: // lartpc-docdb. fnal. gov/ 0000/ 000059/ 001/ stycas2850. pdf*

[Emeb]   EMERSON & CUMING. *Technical Tip: Gamma Irradiation. In: http: // www. lindberg-lund. fi/ files/ Tekniske% 20datablad/ EC-GAMMA-TD. pdf*

[ENE]   ENEA. *Conductor in conduit coils (CICC) of the ITER magnet system, courtesy of ENEA superconducting lab*

[Equ78]   EQUIPE TFR: Tokamak plasma diagnostics. In: *Nuclear fusion* 18 (1978), 5, S. 647

[Eva10]   EVANS, D: Swelling and Gas Evolution in Irradiated Organic Matrix Composites - a Review. In: *Advances in Cryogenic Engineering* 103 (2010), 2010, S. 103–110. ISBN 0735407657

[FBBT12]   FLEITER, J ; BALLARINO, A ; BOTTURA, L ; TIXADOR, P: Electrical characterization of RE-123 Roebel cables. In: *Applied Superconductivity Conference.* Portland : to be published at SuST, 2012

[FHG08]   FRANK, A ; HELLER, R ; GOLDACKER, W ; KLING, A ; SCHMIDT, C: Roebel assembled coated conductor cables (RACC): Ac-Losses and current carrying potential. In: *Journal of Physics: Conference Series* 97 (2008), Februar, S. 012147. – ISSN 1742–6596

[FHKa]   FLÜKIGER, R ; HARIHARAN, S Y. ; KÜNTZLER, R ; LUO, H L. ; WEISS, F ; WOLF, T ; XU, J Q.: Nb-Ti. In: FLÜKIGER, R (Hrsg.) ; KLOSE, W (Hrsg.): *Nb based alloys and compounds.* SpringerMaterials - The Landolt-Börnstein Database

[FHKb]   FLÜKIGER, R ; HARIHARAN, S Y. ; KÜNTZLER, R ; LUO, H L. ; WEISS, F ; WOLF, T ; XU, J Q.: Nb$_3$Sn, bulk. In: FLÜKIGER, R (Hrsg.) ; KLOSE, W (Hrsg.):

Nb based alloys and compounds. SpringerMaterials - The Landolt-Börnstein Database

[FRRS08]   FLESHLER, S ; ROY, D ; RUPICH, M W. ; SANTAMARIA, A: Reducing the Price-Performance Ratio of 2G Wire. In: *Internation Workshop on Coated Conductors for Applications*. Huston, 2008

[Fuj]   FUJIKURA. *Fujikura Develops 800-Meter Class High-Performance YBCO Superconducting Wire. In: http://www.fujikura.co.jp/eng/f-news/2034416_4207.html*

[FZZ09]   FENG, K M. ; ZHANG, G S. ; ZHENG, G Y. ; ZHAO, Z ; YUAN, T ; LI, Z Q. ; SHENG, G Z. ; PAN, C H.: Conceptual design study of fusion DEMO plant at SWIP. In: *Fusion Engineering and Design* 84 (2009), Dezember, 12, S. 2109–2113. – ISSN 09203796

[GAS06]   GRILLI, F. ; ASHWORTH, S. P. ; STAVREV, S.: Magnetization AC losses of stacks of YBCO coated conductors. In: *Physica C: Superconductivity* 434 (2006), Februar, 2, S. 185–190. – ISSN 09214534

[Gau01]   GAUCKLER, L J. *Funktionskeramik: Ingenieurkeramik III*. 2001

[GDM04]   GODEKE, A ; DHALLE, M ; MORELLI, A ; STOBBELAAR, L ; VAN WEEREN, H ; VAN ECK, H J N. ; ABBAS, W ; NIJHUIS, A ; DEN OUDEN, A ; TEN HAKEN, B: A device to investigate the axial strain dependence of the critical current density in superconductors. In: *Review of Scientific Instruments* 75 (2004), 12, S. 5112. – ISSN 00346748

[GFH07]   GOLDACKER, W ; FRANK, A ; HELLER, R ; SCHLACHTER, S I. ; RINGSDORF, B ; WEISS, K-P ; SCHMIDT, C ; SCHULLER, S: ROEBEL Assembled Coated Conductors (RACC): Preparation, Properties and Progress. In: *IEEE Transactions on Applied Superconductivity* 17 (2007), Juni, 2, S. 3398–3401. – ISSN 1051–8223

[GFK09]   GOLDACKER, W ; FRANK, A ; KUDYMOW, A ; HELLER, R ; KLING, A ; TERZIEVA, S ; SCHMIDT, C: Status of high transport current ROEBEL assembled coated conductor cables. In: *Superconductor Science and Technology* 22 (2009), März, 3, S. 034003. – ISSN 0953–2048

[GFS03]   GÖMÖRY, F ; FROLEK, L ; SOUC, J ; COLETTA, G ; SPREAFICO, S: Measurement of DC Critical Current inSuperconducting Cable With Non-Uniformities. In: *IEEE Transactions on Applied Superconductivity* 13 (2003), 2, S. 1968–1971

[GGA08]   GARCIA, J ; GIRUZZI, G ; ARTAUD, J F. ; BASIUK, V ; DECKER, J ; IMBEAUX, F ; PEYSSON, Y ; SCHNEIDER, M: Analysis of DEMO scenarios with the CRONOS suite of codes. In: *Nuclear Fusion* 48 (2008), Juli, 7, S. 075007. – ISSN 0029–5515

[Gir06]   GIRUZZI, G: TW5-TRP-005 : HCLL Blanket design from PPCS-model AB / EFDA. 2006. – Forschungsbericht. – 271–280 S

[GKG09]   GUOYAO, Z ; KAIMING, F ; GUANGZHAO, S: Simulation of Plasma Parameters for HCSB-DEMO by a 1.5D Plasma Transport Code. In: *Plasma Science and Technology* 11 (2009), Dezember, 6, S. 651–656. – ISSN 1009–0630

[GKK01]   GODEKE, A ; KROOSHOOP, H J G. ; KNOOPERS, H G. ; TEN HAKEN, B ; TEN KATE, H H J.: Experimental Verification of the Temperature and Strain Dependence of the Critical Properties in Nb_3Sn Wires. In: *IEEE Transactions on Applied Superconductivity* 11 (2001), 1, S. 1526–1529

[GMP04]   GONZÁLEZ, J ; MESTRES, N ; PUIG, T ; GÁZQUEZ, J ; SANDIUMENGE, F ; OBRADORS, X ; USOSKIN, A ; JOOSS, C ; FREYHARDT, H ; FEENSTRA, R: Biaxial texture analysis of $YBa_2Cu_3O_{7-x}$ coated conductors by micro-Raman spectroscopy. In: *Physical Review B* 70 (2004), September, 9, S. 1–8. – ISSN 1098–0121

[GNK06]   GOLDACKER, W ; NAST, R ; KOTZYBA, G ; SCHLACHTER, S I. ; FRANK, a ; RINGSDORF, B ; SCHMIDT, C ; KOMAREK, P: High current DyBCO-ROEBEL Assembled Coated Conductor (RACC). In: *Journal of Physics: Conference Series* 43 (2006), Juni, S. 901–904. – ISSN 1742–6588

[God06]   GODEKE, A: A review of the properties of Nb_3Sn and their variation with A15 composition, morphology and strain state. In: *RF Superconductivity Workshop*. Padua : Lawrence Berkeley National Laboratory, 2006

[Gol11]   GOLDACKER, W: REBCO Cables. In: *HTS4Fusion Conductor Workshop*. KIT CN, Karlsruhe, 2011

[GR74]   GARRETT, K W. ; ROSENBERG, H M.: The thermal conductivity of epoxy-resin/powder composite materials. In: *Journal of Physics D: Applied Physics* 7 (1974), 9, S. 1247–1258

[GSB11]   GLASSON, N ; STAINES, M ; BUCKLEY, R ; PANNU, M ; KALSI, S S.: Development of a 1 MVA 3-Phase Superconducting Transformer Using YBCO Roebel Cable. In: *Applied Superconductivity, IEEE Transactions on* 21 (2011), 99, S. 1–4

[GSR08]   GOLDACKER, W ; SCHLACHTER, S I. ; RINGSDORF, B ; SCHMIDT, C ; WEISS, K-P ; SCHWARZ, M ; FRANK, A ; LAMPE, A ; KLING, A ; HELLER, R: Final Report on HTS Material for Fusion Magnets: EFDA-No. TW5-TMS-HTSPER / Forschungszentrum Karlsruhe GmbH (FZK). Eggenstein-Leopoldshafen, 2008. – Forschungsbericht

[Haw06]   HAWSEY, R A.: HTS 2G Coated Conductor R&D in Japan: Status, Directions and Investments. In: *Wire Developement Workshop*. St. Peterburg, 2006

[Hay]   HAYNES INTERNATIONAL. *Hastelloy alloy information. In: http://www. haynesintl.com/HASTELLOYXAlloy/HastelloyXAlloyPP.htm*

[Haz10]   HAZELTON, D W.: Progress in Coated Conductor at SuperPower. In: *Low Temperature High Field Superconductor Workshop*. Monterey : SuperPower, 2010

[Haz12]   HAZELTON, D W.: 2G HTS Conductors at SuperPower. In: *Low Temperature High Field Superconductor Workshop*. Napa : SuperPower, 2012

[HB01]   HAMACHER, T ; BRADSHAW, A M.: Fusion as a Future Power Source: Recent Achievements and Prospects. In: *18th World Energy Congress*. Buenos Aires : Max-Plack-Institute für Plasmaphysik, 2001

[HCD03]   HELLER, R. ; CIAZYNSKI, D. ; DUCHATEAU, J.L. ; MARCHESE, V. ; SAVOLDI-RICHARD, L. ; ZANINO, R.: Evaluation of the current sharing temperature of the ITER toroidal field model coil. In: *IEEE Transactions on Applied Superconductivity* 13 (2003), Juni, 2, S. 1447–1451. – ISSN 1051–8223

[HD11]   HELLER, R ; DUCHATEAU, J L.: Which Superconducting Magnets for DEMO and Future Fusion Reactors? In: *HTS4Fusion Conductor Workshop*. KIT CN, Karlsruhe, 2011

[Hel07]   HELLER, R: Final Report on Scoping Study of HTS Fusion Magnets - EFDA-No: TW5.TMS-HTSMAG Deliverable 6 / FZK - ITEP. Karlsruhe, 2007. – Forschungsbericht

[Hel09]   HELLER, R: HTS-Stack Weichlötung Test mit Original-Bandstapel / Karlsruhe Institute of Technology. Eggenstein-Leopoldshafen, 2009. – Forschungsbericht

[Hen78]   HENNING, C D.: Superconductivity for Mirror Fusion. In: *Applied Superconductivity Conference*. Pittsburg, 1978

[HFPZ06]   HONG, S ; FIELD, M B. ; PARRELL, J A. ; ZHANG, Y: Latest Improvements of Current Carrying Capability of Niobium Tin and Its Magnet Applications. In:

IEEE Transactions on Applied Superconductivity 16 (2006), Juni, 2, S. 1146–1151. – ISSN 1051–8223

[HLCB82]   HAUBENREICH, P N. ; LUBELL, M S. ; CORNISH, D N. ; BEARD, D S.: Superconducting magnets for fusion. In: *Nuclear Fusion* 22 (1982), 9, S. 1209–1236

[HM75]   HOENIG, M O. ; MONTGOMERY, D B.: Dense Supercritical-Helium Cooled Superconductors for Large High Field Stabilized Magnets. In: *IEEE Transactions on Magnetics* 11 (1975), 2, S. 2–5

[HNC71]   HENNING, C D. ; NELSON, R L. ; CHARGIN, A K. ; DENHOY, B S. ; HARSHBARGER, F: Engineering the Baseball Magnet System. In: *Nuclear Science, IEEE Transactions on* 18 (1971), 4, S. 290–295

[HOA05]   HIWATARI, R ; OKANO, K ; ASAOKA, Y ; SHINYA, K ; OGAWA, Y: Demonstration tokamak fusion power plant for early realization of net electric power generation. In: *Nuclear Fusion* 45 (2005), Februar, 2, S. 96–109. – ISSN 0029–5515

[HOAO07]   HIWATARI, R ; OKANO, K ; ASAOKA, Y ; OGAWA, Y: Analysis of critical development issues towards advanced tokamak power plant CREST. In: *Nuclear Fusion* 47 (2007), Mai, 5, S. 387–394. – ISSN 0029–5515

[HP]   HP ALLOYS. *NITRONIC® 40 (S21900 / S21904)*. In: `http://www.hpalloy.com/alloys/descriptions/NITRONIC40.html`

[HRB11]   HAZELTON, D W. ; ROY, F ; BROWNSEY, P: Recent Developments in 2G HTS Coil Technology. In: *European Conference on Applied Superconductivity*. Den Haag : SuperPower, 2011

[Hul03]   HULL, J R.: Applications of high temperature superconductors in power technology. In: *Reports on Progress in Physics* 66 (2003), 11

[IBK10]   ISHIDA, S ; BARABASCHI, P ; KAMADA, Y: Status and prospect of the JT-60SA project. In: *Fusion Engineering and Design* 85 (2010), Dezember, 10-12, S. 2070–2079. – ISSN 09203796

[IM91]   IVANENKO, O M. ; MITSEN, K V.: Low temperature direct measurements of H_{c2} in HTSC using megagauss magnetic fields. In: *Physica C: Superconductivity* 185-189 (1991), 3, S. 1859–1860

[IMH10]   ISHIDA, M ; MAEKI, K ; HIWATARI, R ; BONNIN, X ; ZHU, S ; HATAYAMA, A ; SCHNEIDER, R ; COSTER, D: Numerical Analysis of Divertor Plasma for

Demo-CREST. In: *Contributions to Plasma Physics* 50 (2010), Mai, 3-5, S. 362–367. – ISSN 08631042

[ITE] ITER. *Magnets. In:* $http://www.iter.org/mach/magnets$

[ITE09a] ITER: Design Description Document: DDD 11 - Magnets - 1. Design Basics (ITER-D-2NPLKM v1.8). 2009. – Forschungsbericht

[ITE09b] ITER: Design Description Document: DDD 11 - Magnets - 2. TF Coils and Structures (ITER-D-2MVZNX v2.2). 2009. – Forschungsbericht

[ITE09c] ITER: Design Description Document: DDD 11 - Magnets - 3. CS Coils and Pre-Compression Structure (ITER-D-2NHKHH v1.5). 2009. – Forschungsbericht

[ITE09d] ITER: Design Description Document: DDD 11 - Magnets - 4. PF Coils and Supports (ITER-D-2LGJUP v3.0). 2009. – Forschungsbericht

[ITE09e] ITER: Design Description Document: DDD 11 - Magnets - 7. Conductors (ITER-D-2NBKXY v1.2). 2009. – Forschungsbericht

[ITE10] ITER. *ITER magnet system. In:* $http://www.iter.org/$. Dezember 2010

[Iwa09] IWASA, Y: *Case studies in superconducting magnets.* 2. New York : Springer, Oktober 2009. – 885 S. – ISBN 978–0–387–09799–2

[JIN97] JOTAKI, E ; ITOH, S ; NAKAMURA, K: Impurity accumulation in a Nb_3Sn superconducting magnet system on the high-field tokamak TRIAM-1M. In: *Fusion Engineering and Design* 36 (1997), S. 289–297

[JKS09] JONG, C ; KNASTER, J ; SBORCHIA, C: The Electrical Insulation of the TF Coils of ITER. In: *Fusion Science and Technology* 56 (2009), 2, S. 666–671

[JMH13] JIANG, J ; MIAO, H ; HUANG, Y ; HONG, S ; PARRELL, J A. ; SCHEUERLEIN, C ; DI MICHIEL, M ; GHOSH, A K. ; TROCIEWITZ, U P. ; HELLSTROM, E E. ; LARBALESTIER, D C.: Reduction of Gas Bubbles and Improved Critical Current Density in Bi-2212 Round Wire by Swaging. In: *IEEE Transactions on Applied Superconductivity* 23 (2013), Juni, 3, S. 6400206–6400206. – ISSN 1051–8223

[Jos99] JOSEPH, G: *COPPER Its Trade, Manufacture, Use and Enviromental Status.* 1. Materials Park : ASM International, 1999. – ISBN 0–87170–656–3

[Kar12] KARIO, A: Investigation of the Coated Conductors Rutherford-cable using ROEBEL-cables as strands. In: *Applied Superconductivity Conference.* Portland : to be published at SuST, 2012

[KBB08]   KOLBASOV, B ; BELYAKOV, V ; BONDARCHUK, E ; BORISOV, a ; KIRILLOV, I ; LEONOV, V ; SHATALOV, G ; SOKOLOV, Y ; STREBKOV, Y ; VASILIEV, N: Russian concept for a DEMO-S demonstration fusion power reactor. In: *Fusion Engineering and Design* 83 (2008), Dezember, 7-9, S. 870–876. – ISSN 09203796

[Kel08]   KELLER, P: *Thermal and electromechnical investigations on Bi-2223 based technical superconducting tapes*, Ruprecht-Karls-Universität Heidelberg, Diploma Thesis, 2008

[Kem12a]   KEMP, R: Demo Design Summary - March 2012 - (EFDA-D-2L2F7V v1.0) / CCFE. 2012. – Forschungsbericht

[Kem12b]   KEMP, R. *PROCESS - Power Reactor Optimisation Code - Version 3029*. 2012

[KFC06]   KRABBES, G ; FUCHS, G ; CANDERS, W-R ; MAY, H ; PALKA, R: *High Temperature Superconductor Bulk Materials*. Weinheim : WILEY-VCH, 2006. – ISBN 978–3–527–40383–7

[KGSI06]   KALSI, S S. ; GAMBLE, B B. ; SNITCHLER, G ; IGE, S O.: The status of HTS ship propulsion motor developments. In: *IEEE Power Engineering Society General Meeting* (2006), S. 5 pp. ISBN 1–4244–0493–2

[KHS63]   KIM, Y B. ; HEMPDTEAD, C F. ; STRNAD, A R.: Magnetization and Critical Supercurrents. In: *Phys. Rev. Lett.* 129 (1963), 2, S. 528–535

[KKS01]   KASABA, K ; KATAGIRI, K ; SHOJI, Y ; TAKAHASHI, T ; NOTO, K ; GOTO, K ; SAITO, T ; KONO, O: Stress-strain dependence of critical current in Nb_3Sn superconducting wires stabilized with Cu-Nb microcomposites - effect of Nb content. In: *Cryogenics* 41 (2001), Januar, 1, S. 9–14. – ISSN 00112275

[Kom95]   KOMAREK, P: *Hochstromanwendungen der Supraleitung*. 1. Stuttgart : B.G.Teubner, 1995

[KPP05]   KIM, K. ; PARK, HK ; PARK, KR ; LIM, BS ; LEE, SI ; CHU, Y ; CHUNG, WH ; OH, YK ; BAEK, SH ; LEE, SJ ; OTHERS: Status of the KSTAR superconducting magnet system development. In: *Nuclear fusion* 45 (2005), S. 783–789

[KSO91]   KIKUCHI, M ; SEKI, Y ; OIKAWA, A ; ANDO, T ; OHARA, Y ; NISHIO, S ; SEKI, M ; TAKIZUKA, T ; TANI, K ; OZEKI, T: Conceptual design of the steady state tokamak reactor (SSTR). In: *Fusion Engineering and Design* 18 (1991), Dezember, S. 195–202. – ISSN 09203796

[KSSP00]   KRIVCHENKOV, Y ; SBORCHIA, C ; STEPANOV, B ; PANIN, A: Assessment of the radial plate design of the ITER TF coil winding pack. In: *Applied Superconductivity, IEEE Transactions on* 10 (2000), März, 1, S. 592–595. – ISSN 10518223

[Kud12]   KUDYMOW, A: Roebel Assembled Coated Conductor (RACC) cable contacts. In: *personal communication* (2012)

[Laa09]   VAN DER LAAN, D C.: YBa2Cu3O7-delta coated conductor cabling for low ac-loss and high-field magnet applications. In: *Superconductor Science and Technology* 22 (2009), Juni, 6, S. 065013. – ISSN 0953–2048

[Laa12]   VAN DER LAAN, D C.: High-temperature superconducting Conductor on Round Core (CORC) cables for power transmission and magnet applications. In: *ITEP - Institute Seminar.* KIT CN, Karlsruhe : KIT CN, 2012

[Laka]   LAKESHORE. *Cernox. In:* http: // www. lakeshore. com/ products/ Cryogenic-Temperature-Sensors/ Cernox/ Models/ Pages/ Overview. aspx

[Lakb]   LAKESHORE. *DT-670 Silicon Diodes. In:* http: // www. lakeshore. com/ products/ Cryogenic-Temperature-Sensors/ Silicon-Diodes/ DT-670/ Pages/ Overview. aspx

[Lakc]   LAKESHORE. *Typical sensor performance. In:* http: // www. lakeshore. com/ products/ Cryogenic-Temperature-Monitors/ Model-218/ Pages/ Typical-Sensor-Performance. aspx

[Lam06]   LAMPE, A: *Elektro-mechanische Charakterisierung von technischen Supraleitern bei kryogenen Temperaturen*, Technische Hochschule Wildau, Diploma Thesis, 2006

[LAP11]   VAN DER LAAN, D C. ; ABRAIMOV, D ; POLYANSKII, A A. ; LARBALESTIER, D C. ; DOUGLAS, J F. ; SEMERAD, R ; BAUER, M: Anisotropic in-plane reversible strain effect in $Y_{0.5}$ $Gd_{0.5}Ba_2Cu_3O_{7-\delta}$ coated conductors. In: *Superconductor Science and Technology* 24 (2011), November, 11, S. 115010. – ISSN 0953–2048

[LCD05]   LIBEYRE, P ; CIAZYNSKI, D ; DUCHATEAU, J ; SCHILD, T ; FIETZ, W ; ZAHN, G: How should we test the ITER TF coils? In: *Fusion Engineering and Design* 75-79 (2005), November, S. 81–85. – ISSN 09203796

[LCZ08]   LU, J ; CHOI, E S. ; ZHOU, H D.: Physical properties of Hastelloy® C-276™at cryogenic temperatures. In: *Journal of Applied Physics* 103 (2008), 6, S. 064908. – ISSN 00218979

[LECS07]   VAN DER LAAN, D C. ; EKIN, J W. ; CLICKNER, C C. ; STAUFFER, T C.: Delamination strength of YBCO coated conductors under transverse tensile stress. In: *Superconductor Science and Technology* 20 (2007), August, 8, S. 765–770. – ISSN 0953–2048

[Lee01]   LEE, P J.: *Engineering superconductivity*. New York : Wiley-Interscience, 2001. – ISBN 0471411167

[Lee11]   LEE, P J. *Simplified Engineering Critical Current Density vs. Applied Field. In: http: // magnet. fsu. edu/ ˜lee/plot/plot. htm*. 2011

[LEH01]   VAN DER LAAN, D C. ; VAN ECK, H J N. ; TEN HAKEN, B ; SCHWARTZ, J ; TEN KATE, H H j.: Temperature and magnetic field dependence of the critical current of $Bi_2Sr_2Ca_2Cu_3O_X$ tape conductors conductor. In: *IEEE Transactions on Applied Superconductivity* 11 (2001), I, S. 3345–3348

[Leh13]   LEHNER, T F.: Production of 2G HTS Conductor at SuperPower: Recent Progress and Ongoing Improvements. In: *MEM13 Workshop*. Aix en Provence : Super-Power, 2013

[Liu11]   LIU, X: Improved methods for heat treating Bi2Sr2CaCu2Ox/Ag round wires and cables. In: *22nd International Conference on Magnet Technology*, 2011

[LLG11]   VAN DER LAAN, D C. ; LU, X F. ; GOODRICH, L F.: Compact GdBa 2 Cu 3 O 7-delta coated conductor cables for electric power transmission and magnet applications. In: *Superconductor Science and Technology* 24 (2011), April, 4, S. 042001. – ISSN 0953–2048

[LLG13]   VAN DER LAAN, D C. ; LU, X F. ; GOODRICH, L F. ; HAUGAN, T J. ; CLICKNER, C C. ; STAUFFER, T C. ; ABRAIMOV, D ; POLYANSKII, A A. ; MILLER, G E. ; NOYES, P D. ; WEIJERS, H W. ; LARBALESTIER, D C. ; SEMERAD, R ; BAUER, M ; BROMBERG, L ; MINERVINI, J V.: Anisotropic reversible strain effect in REBCO. In: *MEM13 Workshop*. Aix en Provence, 2013

[LMB09]   LIBEYRE, P ; MITCHELL, N ; BESSETTE, D ; GRIBOV, Y ; JONG, C ; LYRAUD, C: Detailed design of the ITER central solenoid. In: *Fusion Engineering and Design* 84 (2009), Juni, 7-11, S. 1188–1191. – ISSN 09203796

[LMS02]   LINDAU, R ; MÖSLANG, A ; SCHIRRA, M: Thermal and mechanical behaviour of the reduced-activation- ferritic-martensitic steel EUROFER. In: *Fusion Engineering and Design* 61-62 (2002), S. 659–664

[Lon11]   LONG, N J.: HTS Roebel cables. In: *HTS4Fusion Conductor Workshop*. KIT CN, Karlsruhe, 2011

[Lon12]   LONG, N J.: HTS Roebel cables. In: *ITEP - Institute Seminar*. KIT CN, Karlsruhe, 2012

[Lor11]   LORD, E: SuperPower YBCO price and performance. In: *personal communication* (2011)

[LPLL06]   LEE, P J. ; PARK, M ; LIM, H ; LEE, H G.: Magnetization Loss and Shield Effect in Multi-Stacked Tapes With Various Stacking Configurations. In: *Applied Superconductivity, IEEE Transactions on* 16 (2006), 2, S. 131–134

[LRT09]   LI, X ; RUPICH, MW ; THIEME, C L H. ; TEPLITSKY, M ; SATHYAMURTHY, S ; THOMPSON, E ; BUCZEK, D ; SCHREIBER, J ; DEMORANVILLE, K ; LYNCH, J ; INCH, J ; TUCKER, D ; SAVOY, R ; FLESHLER, S: The Development of Second Generation HTS Wire at American Superconductor. In: *Applied Superconductivity, IEEE Transactions on* 19 (2009), 3, S. 3231–3235

[LSB10]   LAKSHMI, L S. ; STAINES, M P. ; BADCOCK, R a. ; LONG, N J. ; MAJOROS, M ; COLLINGS, E W. ; SUMPTION, M D.: Frequency dependence of magnetic ac loss in a Roebel cable made of YBCO on a Ni-W substrate. In: *Superconductor Science and Technology* 23 (2010), August, 8, S. 085009. – ISSN 0953–2048

[LSI11]   LIM, B ; SIMON, F ; ILYIN, Y ; GUNG, C Y. ; SMITH, J ; HSU, Y H. ; LUONGO, C ; JONG, C ; MITCHELL, N: Design of the ITER PF Coils. In: *Applied Superconductivity, IEEE Transactions on* 21 (2011), 3, S. 1918–1921

[Luc69]   LUCAS, E. J.: Results of Tests on an 8.8 Tesla, 51-cm-Bore Magnet System. In: *Journal of Applied Physics* 40 (1969), 5, S. 2101–2106. – ISSN 00218979

[LWB11]   LIU, B ; WU, Y ; BRUZZONE, P: The results of the second Chinese TF conductor sample. In: *Fusion Engineering and Design* 86 (2011), Juni, 4-5, S. 369–372. – ISSN 09203796

[LZL08]   LI, J F. ; ZHANG, P X. ; LIU, X H. ; LI, J S. ; FENG, Y ; DU, S J. ; WANG, T C. ; LIU, W T. ; GRUNBLATT, G ; VERWAERDE, C ; HOANG, G K.: The microstructure of NbTi superconducting composite wire for ITER project. In:

Physica C: Superconductivity 468 (2008), September, 15-20, S. 1840–1842. – ISSN 09214534

[Mac11] MACKOWIAK, B: Knappe Ressourcen - Die Quellen des Kältetreibstoffs Helium versiegen. In: *Zeit Online* (2011), Oktober, S. 1–2

[Mae12] MAEDER, O: *Simulation und Experimente zur Stabilität von HTSL-Bandleitern*, KIT, Ph.D. - thesis, 2012

[Max12a] MAX-PLANCK-INSTITUT FÜR PLASMAPHYSIK. *Stellarator magnet system. In: http://www.dpg-physik.de/dpg/gliederung/fv/p/info/pix/Stellarator2.jpg.* 2012

[Max12b] MAX-PLANCK-INSTITUT FÜR PLASMAPHYSIK. *Tokamak magnet system. In: http://www.dpg-physik.de/dpg/gliederung/fv/p/info/pix/Tokamak.jpg.* 2012

[MBB09] MITCHELL, N ; BAUER, P ; BESSETTE, D ; DEVRED, A ; GALLIX, R ; JONG, C ; KNASTER, J ; LIBEYRE, P ; LIM, B ; SAHU, A: Status of the ITER magnets. In: *Fusion Engineering and Design* 84 (2009), Juni, 2-6, S. 113–121. – ISSN 09203796

[MBT11] MINERVINI, J V. ; BROMBERG, L ; TAKAYASU, M: HTS and Cable Concepts. In: *HTS4Fusion Conductor Workshop.* KIT CN, Karlsruhe, 2011

[MCP05] MAISONNIER, D ; COOK, I ; PIERRE, S ; LORENZO, B ; EDGAR, B ; KARIN, B ; LUIGI, D ; ROBIN, F ; LUCIANO, G ; STEPHAN, H: The European power plant conceptual study. In: *Fusion Engineering and Design* 75-79 (2005), November, S. 1173–1179. – ISSN 09203796

[MCP06] MAISONNIER, D ; COOK, I ; PIERRE, S ; LORENZO, B ; LUIGI, D ; LUCIANO, G ; PRACHAI, N ; ALDO, P: DEMO and fusion power plant conceptual studies in Europe. In: *Fusion Engineering and Design* 81 (2006), Februar, 8-14, S. 1123–1130. – ISSN 09203796

[MCS05] MAISONNIER, D ; COOK, I ; SARDAIN, P ; ANDREANI, R ; DI PACE, L ; FORREST, R ; GIANCARLI, L ; HERMSMEYER, S ; NORAJITRA, P ; TAYLOR, N ; WARD, D J.: A conceptual study of commercial fusion power plants / EFDA. 2005 (05). – Forschungsbericht. – 1–212 S

[MGGC54] MATTHIAS, B T. ; GEBALLE, T H. ; GELLER, S ; CORENZWIT, E: Superconductivity of Nb$_3$Sn. In: *Physical Review* 95 (1954), 6, S. 1954

[MKJ11]   MALAGOLI, A ; KAMETANI, F ; JIANG, J ; TROCIEWITZ, U P. ; HELLSTROM, E E. ; LARBALESTIER, D C.: Evidence for long range movement of Bi-2212 within the filament bundle on melting and its significant effect on Jc. In: *Superconductor Science and Technology* 24 (2011), Juli, 7, S. 075016. – ISSN 0953–2048

[MKT93]   MURASE, S ; KOBATAKE, S ; TANAKA, M ; TASHIRO, I ; HORIGAMI, O ; OGIWARA, H ; SHIBATA, K ; NAGAI, K ; ISHIKAWA, K: Effects of a high magnetic field on fracture toughness at 4.2 K for austenitic stainless steels. In: *Fusion Engineering and Design* 20 (1993), Januar, S. 451–454. – ISSN 09203796

[MMH09]   MANN, T L. ; MARTOVETSKY, N ; HATFIELD, D ; KENNEY, S ; MILLER, J ; KYNOCH, J ; BAI, H ; BIRD, M: Cold test facility for the ITER central solenoid coils. In: *23rd IEEE/NPSS Symposium on Fusion Engineering* (2009), Juni, S. 1–4. ISBN 978–1–4244–2635–5

[MMM03]   MARKEN, K R. ; MIAO, H ; MEINESZ, M ; CZABAJ, B ; HONG, S: BSCCO-2212 conductor development at oxford superconducting technology. In: *IEEE Transactions on Applied Superconductivity* 13 (2003), Juni, 2, S. 3335–3338. – ISSN 1051–8223

[MO12]   MEISSNER, W ; OCHSENFELD, R: Ein neuer Effekt bei Eintritt der Supraleit-fähigkeit. In: *Die Naturwissenschaften* 21 (1912), 44

[Mos00]   MOSHCHALKOV, V V.: Nanostructured Superconductors: How to enhance Vortex Pinning and Critical Fields? In: *Boulder 2000 Summer School for Condensed Matter and Materials Physics.* Boulder, 2000

[MPC03]   MA, K. B. ; POSTREKHIN, Y. V. ; CHU, W. K.: Superconductor and magnet levitation devices. In: *Review of Scientific Instruments* 74 (2003), 12, S. 4989. – ISSN 00346748

[MPV12]   MARTIN, E ; PAVIA, F ; VENTRELLA, A ; AVALLE, M ; LARA-CURZIO, E ; FERRARIS, M: A Brazilian Disk Test for the Evaluation of the Shear Strength of Epoxy-Joined Ceramics. In: *International Journal of Applied Ceramic Technology* 9 (2012), Juli, 4, S. 808–815. – ISSN 1546542X

[MS08]   MÜLLER, H ; SCHNEIDER, T: Heat treatment of Nb_3Sn conductors. In: *Cryogenics* 48 (2008), Juli, 7-8, S. 323–330. – ISSN 00112275

[MSI06a]   MITO, T ; SAGARA, A ; IMAGAWA, S ; YAMADA, S ; TAKAHATA, K ; YANAGI, N ; CHIKARAISHI, H ; MAEKAWA, R ; IWAMOTO, A ; HAMAGUCHI, S: Applied superconductivity and cryogenic research activities in NIFS. In: *Fusion Engineering and Design* 81 (2006), November, 20-22, S. 2389–2400. – ISSN 09203796

[MSI06b] MOTOJIMA, O ; SAKAKIBARA, S ; IMAGAWA, S ; SAGARA, A ; SEKI, T ; MUTOH, T ; MORISAKI, T ; KOMORI, A ; OHYABU, N ; YAMADA, H: Progress of plasma experiments and superconducting technology in LHD. In: *Fusion Engineering and Design* 81 (2006), November, 20-22, S. 2277–2286. – ISSN 09203796

[MYI07] MUKOYAMA, S ; YAGI, M ; ICHIKAWA, M ; TORII, S ; TAKAHASHI, T ; SUZUKI, H ; YASUDA, K: Experimental Results of a 500 m HTS Power Cable Field Test. In: *IEEE Transactions on Applied Superconductivity* 17 (2007), Juni, 2, S. 1680–1683. – ISSN 1051–8223

[NAB06] NAJMABADI, F ; ABDOU, A ; BROMBERG, L ; BROWN, T ; CHAN, V ; CHU, M ; DAHLGREN, F ; ELGUEBALY, L ; HEITZENROEDER, P ; HENDERSON, D: The ARIES-AT advanced tokamak, Advanced technology fusion power plant. In: *Fusion Engineering and Design* 80 (2006), Januar, 1-4, S. 3–23. – ISSN 09203796

[Nak88] NAKAJIMA, H: Design of the Nb3Sn Demo Poloidal Coil (DPC-EX). In: *IEEE Transactions on Magnetics* 24 (1988), 2, S. 1444–1447

[NIS03] NUNOYA, Y ; ISONO, T ; SUGIMOTO, M ; TAKAHASHI, Y ; NISHIJIMA, G ; MATSUI, K ; KOIZUMI, N ; ANDO, T ; OKUNO, K: Evaluation Method of Critical Current and Current Sharing Temperature for Large-Current Cable-in-Conduit Conductors. In: *IEEE Transactions on Applied Superconductivity* 13 (2003), Juni, 2, S. 1404–1407. – ISSN 1051–8223

[NJW99] NOE, M ; JUENGST, K-P ; WERFEL, F N. ; BAECKER, M ; BOCK, J ; ELSCHNER, S: Quench behaviour of high temperature superconducting bulk material in resistive fault current limiters. In: *4th European Conference on Applied Superconductivity*. Sitges, 1999

[NNM01] NAGAMATSU, J ; NAKAGAWA, N ; MURANAKA, T ; ZENITANI, Y ; AKIMITSU, J: Superconductivity at 39 K in magnesium diboride. In: *Nature* 410 (2001), März, 6824, S. 63–4. – ISSN 0028–0836

[NNY11] NUNOYA, Y ; NABARA, Y ; YOSHIKAWA, M ; MATSUI, K ; HEMMI, T ; TAKAHASHI, Y ; ISONO, T ; KOIUMI, N ; NAKAJIMA, H ; STEPANOV, B ; BRUZZONE, P: Test Result of a Full-Size Conductor Developed for the ITER TF Coils. In: *Applied Superconductivity, IEEE Transactions on* 21 (2011), 3, S. 1982–1986

[NSC77]   NOVIKOV, S I. ; STAVISSKY, B A. ; CHERNOPLEKOV, N A.:   Test Results of "Tokamak-7" Superconducting Magnet System (SMS) Sections. In: *IEEE Transactions on Magnetics* 13 (1977), 1, S. 694–695

[Nyi05]   NYILAS, A: Transducers for Sub-Micron Displacement Measurements at Cryogenic Temperatures. In: *Advances in cryogenic engineering* 824 (2005), S. 27–34

[Oak]   OAK RIDGE NATIONAL LABORATORY. *Fusion Facts. In: http://www.ornl.gov/sci/fed/Fusion-Facts/fusn.htm*

[Oak96]   OAK RIDGE NATIONAL, Laboratory.   *The Chemistry of Superconductors. In: http://www.ornl.gov/info/reports/m/ornlm3063r1/pt5.html.* 1996

[Oh11]   OH, S-S: R&D status of ReBCO coated conductors in Korea.   In: *MEM11 Workshop.* Okinawa, 2011

[Ohs08]   OHSAKI, H: Impact of High-Temperature Superconductors on the Superconducting Maglev / The University of Tokyo, Graduate Shool of Frontier Sciences. Kashiwa, 2008. – Forschungsbericht

[OLP09]   OOMEN, M P. ; LEGHISSA, M ; PROELSS, N: Transposed-Cable Coil & Saddle Coils of HTS for Rotating Machines: Test Results at 30 K. In: *Applied Superconductivity, IEEE Transactions on* 19 (2009), Juni, 3, S. 1633–1638. – ISSN 1051–8223

[Onn12]   ONNES, H K.:   Further experiments with Liquid Helium G. On the electrical resistance of Pure Metals etc. VI. On the Sudden Change in the Rate at which the Resistance of Mercury Disappears. In: *KNAW Proceedings 14 II.* Amsterdam : Huygens Institute - Royal Netherlands Academy of Arts and Sciences, 1912, S. 818–821

[OYM10]   OHYA, M ; YUMURA, H ; MASUDA, T ; NAGAISHI, T ; SHINGAI, Y ; FUJIWARA, N: AC loss characteristics of RE-123 superconducting cable. In: *Journal of Physics: Conference Series* 234 (2010), 3

[PCC81]   PAUL, J W M. ; CLARK, W H M. ; CORDEY, J G. ; FIELDING, S J. ; GILL, R D. ; HUGILL, J ; MCCRACKEN, M ; START, D F H. ; STOTT, P E. ; WOOTTON, A J.: The DITE tokamak experiment. In: *Philosophical Transactions of the Royal Society of London. Series A, Mathematical and Physical Science* 300 (1981), 1456, S. 535–545

[Phi83]   PHILLIPS, J A.: Magnetic Fusion. In: *Winter/Spring Los Alamos Science* (1983)

[PHWM06]   PROKOPEC, R ; HUMER, K ; WEBER, H W. ; MAIX, R K.: IT/P2-11: Assessment of the Mechanical Properties of Candidate ITER Magnet Insulation Systems before and after Reactor Irradiation / Atomic Institute of the Austrian Universities, Vienna University of Technology. Vienna, 2006. – Forschungsbericht

[Pil07]   PILKEY, W D.: *Peterson's stress concentration factors*. 2. Wiley & Sons, Mai 2007. – ISBN 9780471538493

[PNH05]   PRUSSEIT, W ; NEMETSCHEK, R ; HOFFMANN, C ; SIGL, G ; LÜMKEMANN, A ; KINDER, H: ISD process development for coated conductors. In: *Physica C: Superconductivity* 426-431 (2005), Oktober, S. 866–871. – ISSN 09214534

[PPJ07]   PACHER, G W. ; PACHER, H D. ; JANESCHITZ, G ; KUKUSHKIN, A S. ; KOTOV, V ; REITER, D: Modelling of DEMO core plasma consistent with SOL/divertor simulations for long-pulse scenarios with impurity seeding. In: *Nuclear Fusion* 47 (2007), Mai, 5, S. 469–478. – ISSN 0029–5515

[PPL99]   PASTOR, J Y. ; POZA, P ; LLORCA, J: Mechanical Properties of Textured $Bi_2Sr_2CaCu_2O_{8+\delta}$ High-Temperature Superconductors. In: *Journal of the American Ceramic Society* 82 (1999), 11, S. 3139–3144

[PTM04]   PANEK, J S. ; TUTTLE, J G. ; MARRERO, V ; MUSTAFI, S ; EDMONDS, R ; GRAY, A ; RIALL, S: Astro-E2 Magnesium Diboride High Current Leads. In: *AIP Conference Proceedings* 710 (2004), 1, S. 952–957. ISBN 0735401861

[Pul85]   PULSATOR TEAM: The Pulsator tokamak. In: *Nuclear fusion* 25 (1985), 9, S. 1059

[PVGv11]   PARDO, E ; VOJENČIAK, M ; GÖMÖRY, F ; ŠOUC, J: Low-magnetic-field dependence and anisotropy of the critical current density in coated conductors. In: *Superconductor Science and Technology* 24 (2011), Juni, 6, S. 065007. – ISSN 0953–2048

[Quaa]   QUANTUMDESIGN. *Physical Property Measurement System (PPMS®)*. In: *http: //www. qdusa. com/products/ppms. html*

[Quab]   QUANTUMDESIGN. *Thermal Transport Option (TTO)*. In: *http: //www. qdusa. com/sitedocs/productBrochures/tto_rev7-06. pdf*

[Quac]   QUARZWERKE GRUPPE. *Quarzgut: amorph, extren niedrige thermische Ausdehnung. In: http://www. quarzwerke. com/home/maerkte_*

und_ produkte/ Range% 20of% 20products/ englHigh% 20Performance%
20Fillers/ Fused% 20silica

[Quad]   QUARZWERKE GRUPPE. *SILBOND*® *Quarzgut*. In: http://
www. quarzwerke. com/ home/ repository/ Stoffdatenblaetter/
pdf-Dateien/ 3130-SILBOND-QG-de-en. pdf

[Rag11]   RAGHEB, M. *Magnetic confinement fusion*. In: https://netfiles. uiuc.
edu/ mragheb/ www/ NPRE% 20402% 20ME% 20405% 20Nuclear% 20Power%
20Engineering/ Magnetic% 20Confinement% 20Fusion. pdf . 2011

[RBBo10]   RAJAINMAKI, H ; BELLESIA, B ; BONITO-OLIVA, A ; BOTER, E ; BOUTBOUL,
T ; BRATU, E ; CORNELIS, M ; FANTHOME, J ; HARRISON, R ; READMAN, P ;
SBORCHIA, C ; VALENTE, P: ITER Magnets Update - European perspective. In:
Superconductivity in Energy Applications. Tampere, 2010

[Riz12]   RIZZO, E: thermodynamic efficiencies. In: *personal communication* (2012)

[RRG73]   ROTH, J R. ; RICHARDSON, R W. ; GERDIN, G A.: Initial Results from the Nasa
Lewis Bumpy Torus Experiment / Nasa. TMX-71468, 1973. – Forschungsbericht

[RSN73]   REINMANN, J J. ; SWANSON, M C. ; NICHOLS, C R. ; QBLOY, S J. ; NAGY,
L A. ; BRADY, F J.: Nasa Superconducting Magnetic Mirror Facility. In: *Fifth
Symposium on Engineering Problems of Fusion Research*. Princeton : NASA TM
X-71480, 1973

[RWHM92]   ROSNER, C H. ; WALKER, M S. ; HALDAR, P ; MOTOWIDLO, L R.:   Status of
HTS superconductors: Progress in improving transport critical current densities in
HTS Bi-2223 tapes and coils. In: *Cryogenics* 32 (1992), 11, S. 940–948

[SA90]   SMATHERS, D B. ; ALBANY, T W C.: A15 Superconductors. In: *ASM Handbook
Volume 2, Properties and Selection: Nonferrous Alloys and Special-Purpose
Materials*. ASM International, 1990, S. 1060–1076

[Sap00]   SAPPER, J: The superconducting magnet system for the Wendelstein7-X stellarator.
In: *Superconductor Science and Technology* 13 (2000), 5, S. 516–518

[Sax00]   SAXENA, Y C.: Present status of the SST-1 project. In: *Nuclear Fusion* 40 (2000),
6, S. 1069–1082

[SBG01]   SHAFRANOV, V D. ; BONDERENKO, B D. ; GONCHAROV, G A.: On the history
of the research into controlled thermonuclear fusion. In: *Physics-Uspekhi* 44
(2001), August, 8, S. 835–865. – ISSN 1063–7869

[SBOG10]   SAVARY, F ; BONITO-OLIVA, A ; GALLIX, R ; KNASTER, J ; KOIZUMI, N ; MITCHELL, N ; NAKAJIMA, H ; OKUNO, K ; SBORCHIA, C: Status Report on the Toroidal Field Coils for the ITER Project. In: *Applied Superconductivity, IEEE Transactions on* 20 (2010), Juni, 3, S. 381–384. – ISSN 1051–8223

[SBZ12]   SONG, H ; BROWNSEY, P ; ZHANG, Y ; WATERMAN, J ; FUKUSHIMA, T ; HAZELTON, D: 2G HTS Coil Technology Development at SuperPower Inc. In: *Applied Superconductivity Conference.* Portland, 2012

[Sch09]   SCHWARZ, M: *Wärmeleitfähigkeit supraleitender Kompositleiter im Temperaturbereich von 4 K bis 300 K*, Universität Karlsruhe (TH), Ph.D. thesis, 2009

[SD08]   SRINIVASAN, R ; DESHPANDE, S P.: Strategy for the Indian DEMO design. In: *Fusion Engineering and Design* 83 (2008), Dezember, 7-9, S. 889–892. – ISSN 09203796

[SD10]   SELVAMANICKAM, V ; DACKOW, J: Progress in SuperPowerś 2G HTS Wire Development and Manufacturing / 2010 DOE Advanced Cables & Conductors Peer Review. Alexandria, 2010. – Forschungsbericht

[SDR10]   SCHLACHTER, S I. ; DRECHSLER, A ; RINGSDORF, B ; GOLDACKER, W ; GRILLI, F ; HELLER, R: HTS Cables for High-Current Applications. In: *International Conference on Superconductivity and Magnetism.* Antalya, 2010

[See98]   SEEBER, B: Critical current of wires. In: *Handbook of applied superconductivity.* Bristol : Inst. of Physics Publ., 1998. – ISBN 0750303778

[Sel11a]   SELVAMANICKAM, V: 2G HTS Wire for High Magnetic Field Applications 2G HTS wire. In: *HTS4Fusion Conductor Workshop.* KIT CN, Karlsruhe : SuperPower, 2011

[Sel11b]   SELVAMANICKAM, V: Coated Conductors : From R & D to Manufacturing to Commercial Applications Celebrating multiple anniversaries in 2011 ! In: *European Conference on Applied Superconductivity*, SuperPower, 2011

[Sel11c]   SELVAMANICKAM, V: Second-generation HTS Wire for Wind Energy Applications. In: *Symposium on Superconducting Devices for Wind Energy.* Barcelona : SuperPower, 2011

[Sen]   SENGUPTA, A K. *Stress Concentration Caused by Sudden Change in Form.* In: http://web.njit.edu/$\sim$sengupta/met%20301/Stress%20Concentration.pdf

[SFAF07]   SEEBER, B ; FERREIRA, A ; ABÄCHERLI, V ; FLÜKIGER, R: Critical current of a Nb_3Sn bronze route conductor under uniaxial tensile and transverse compressive stress. In: *Superconductor Science and Technology* 20 (2007), September, 9, S. S184–S188. – ISSN 0953–2048

[SGG10]   SCHLACHTER, S I. ; GOLDACKER, W ; GRILLI, F ; HELLER, R ; KUDYMOW, A: Coated Conductor Rutherford Cables (CCRC) for High Current Applications: Concept and Properties. In: *Applied Superconductivity Conference*. Washington DC, 2010

[SGK07]   SCHULLER, S ; GOLDACKER, W ; KLING, A ; KREMPASKY, L ; SCHMIDT, C: Ac-loss measurement of a DyBCO-Roebel assembled coated conductor cable (RACC). In: *Physica C: Superconductivity* 463-465 (2007), Oktober, S. 761–765. – ISSN 09214534

[Sgo11]   SGOBBA, S: Optical observation on 316LN stainless steel tube for ITER TF tube (new ASIPP delivery with nominally better-controlled solution anneal) / CERN - Metallugry and Metrology section. 2011. – Forschungsbericht. – 2–5 S

[SKN93]   SASAKI, T ; KOIZUMI, N ; NISHI, M ; OKUNO, K ; YOSHIDA, K ; TSUJI, H ; MUKAI, H ; WACHI, Y ; HAMAJIMA, T ; NAKAYAMA, S ; FUJIOKA, T: Stability Performance of the DPC-TJ Nb_3Sn Cable-in-Conduit Large Superconducting Coil. In: *Applied Superconductivity, IEEE Transactions on* 3 (1993), 1, S. 523–526

[SKS00]   SHATALOV, G ; KIRILLOV, I ; SOKOLOV, Y ; STREBKOV, Y ; NASILIEV, N: Russian DEMO-S reactor with continuous plasma burn. In: *Fusion Engineering and Design* 51-52 (2000), November, 1-4, S. 289–298. – ISSN 09203796

[SM02]   SATOW, T ; MOTOJIMA, O: Performance of the LHD Magnet System. In: *Applied Superconductivity, IEEE Transactions on* 12 (2002), 1, S. 629–634

[Smi05]   SMITH, C L.: A Fast Track Approach to Fusion Power. In: *CPS*, 2005

[SMM07]   SEKITANI, T ; MATSUDA, Y H. ; MIURA, N: Measurement of the upper critical field of optimally-doped $YBa_2 Cu_3O_{7-\delta}$ in megagauss magnetic fields. In: *New Journal of Physics* 9 (2007), März, 3, S. 47–47. – ISSN 1367–2630

[SMS11]   SUGANO, M ; MACHIYA, S ; SATO, M ; KOGANEZAWA, T ; SHIKIMACHI, K ; HIRANO, N ; NAGAYA, S: Bending strain analysis considering a shift of the neutral axis for YBCO coated conductors with and without a Cu stabilizing layer. In: *Superconductor Science and Technology* 24 (2011), Juli, 7, S. 075019. – ISSN 0953–2048

[Sok94]   SOKOLOV, Y: Study of DEMO reactor concept in Russia. In: *Fusion Engineering and Design* 25 (1994), August, 1-3, S. 159–167. – ISSN 09203796

[Som28]   SOMMERFELD, A: The electron theory of metals on the basis of Fermi statistics. In: *Z. Phys.* 47 (1928), S. 1–32

[SPT91]   STEEVES, M M. ; PAINTER, T A. ; TAKAYASU, M ; RANDALL, R N. ; TRACEY, J E. ; HWANG, L S. ; HOENIG, M O.: The US Demonstration Poloidal Coil. In: *IEEE Transactions on Magnetics* 27 (1991), 2, S. 2369–2372

[SR]   SCHLACHTER, S I. ; RINGSDORF, B: Angle Dependent Electro-Magnetic Measurements of different HTS tapes. In: *to be published*

[Sta12]   STADT MANNHEIM. *Der Roebelstab von Ludwig Roebel (1878 - 1934). In:* http: // www. mannheim. de/ wirtschaft-entwickeln/ roebelstab-ludwig-roebel-1878-1934 . 2012

[Sti98]   STIX, T H.: Highlights in Early Stellarator Research at Princeton. In: *Journal of Plasma Fusion Research* 1 (1998), S. 3–8

[Str12]   STRAUSS, S: influence of pressure on soldering joints. In: *personal communication* (2012)

[Suma]   SUMITOMO ELECTRIC. *Di-BSCCO. In:* http: // global-sei. com/ super/ hts_ e/ index. html

[Sumb]   SUMITOMO ELECTRIC. *Di-BSCCO Type G. In:* http: // global-sei. com/ super/ hts_ e/ type_ g. html

[Sumc]   SUMITOMO ELECTRIC. *Di-BSCCO Type HT. In:* http: // global-sei. com/ super/ hts_ e/ type_ ht. html

[Supa]   SUPERCONDUCTOR, American. *Second generation HTS wire for high current cable and power-dense coil applications. In:* http: // www. amsc. com/ products/ amperiumwire

[SUPb]   SUPERCONDUCTORS.ORG. *Type I Superconductors. In:* http: // www. superconductors. org/ Type1. htm

[SUPc]   SUPERCONDUCTORS.ORG. *Type II Superconductors. In:* http: // www. superconductors. org/ Type2. htm

[Supd]   SUPERPOWER.   *2G HTS Wire Specifications. In: http://www. superpower-inc.com/system/files/SP_2G+Wire+Spec+Sheet_for+ web_0509.pdf*

[Supe]   SUPERPOWER. *Products & Services. In: http://www.superpower-inc. com/content/products-services*

[SXM09]   SELVAMANICKAM, V ; XIE, Y Y. ; MARTCHEVSKI, M ; RAR, A ; SCHMIDT, R M. ; KNOLL, A ; LENSETH, K P. ; WEBER, C S.: High Performance 2G Wires: From R&D to Pilot-Scale Manufacturing. In: *IEEE Transactions on Applied Superconductivity* 19 (2009), Juni, 3, S. 3225–3230. – ISSN 1051–8223

[TA09]   TAVANA, A ; AKHAVAN, M: How T_c can go above 100 K in the YBCO family. In: *The European Physical Journal B* 73 (2009), November, 1, S. 79–83. – ISSN 1434–6028

[Tan06]   TANNA, V L.: *Design and Analysis of the Superconducting Current Feeder System for the International Thermaonuclear Experimental Reactor (ITER)*, University of Karlsruhe, Ph.D. thesis, 2006

[TBL07]   TURRIONI, D ; BARZI, E ; LAMM, M ; LOMBARDO, V ; THIEME, C ; ZLOBIN, A V.: Angular Measurements of HTS Critical Current for High Field Solenoids. In: *Fermilab-Conf*, 2007

[TCBM11]   TAKAYASU, M ; CHIESA, L ; BROMBERG, L ; MINERVINI, J V.: Cabling Method for High Current Conductors Made of HTS Tapes. In: *Applied Superconductivity, IEEE Transactions on* 21 (2011), 3, S. 2340–2344

[TCRM99]   TAKAYASU, M ; CHILDS, R A. ; RANDALL, R N. ; MINERVINI, J V.: ITER Niobium-Tin Strands Reacted under Model Coil Heat-Treatment Conditions. In: *IEEE Transactions on Applied Superconductivity* 9 (1999), 2, S. 644–647

[TH05]   TAYLOR, D M J. ; HAMPSHIRE, D P.: Relationship between the n-value and critical current in Nb_3Sn superconducting wires exhibiting intrinsic and extrinsic behaviour. In: *Superconductor Science and Technology* 18 (2005), 12, S. 297–302

[THT10]   TAKEMATSU, T ; HU, R ; TAKAO, T ; YANAGISAWA, Y ; NAKAGOME, H ; UGLIETTI, D ; KIYOSHI, T ; TAKAHASHI, M ; MAEDA, H: Degradation of the performance of a YBCO-coated conductor double pancake coil due to epoxy impregnation. In: *Physica C: Superconductivity* 470 (2010), September, 17-18, S. 674–677. – ISSN 09214534

[TIM00]   TAKAHATA, K ; IWAMOTO, A ; MAEKAWA, R ; MITO, T ; SATOW, T ; SATOH, S ; MOTOJIMA, O: Hydraulic characteristics of cable-in-conduit conductors for large helical device. In: *Advances in cryogenic engineering* 45 (2000), S. 1111–1118

[TJH10]   TURENNE, M ; JOHNSON, R P. ; HUNTE, F ; SCHWARTZ, J ; SONG, H: Roebel Cable for High-Field Low-Loss Accelerator Magnets. In: *Proceeding of the IPAC'10*. Kyoto, 2010, S. 397–399

[TJS11]   THAKUR, K P. ; JIANG, Z ; STAINES, M P. ; LONG, N J. ; BADCOCK, R A. ; RAJ, A: Current carrying capability of HTS Roebel cable. In: *Physica C: Superconductivity* 471 (2011), Januar, 1-2, S. 42–47. – ISSN 09214534

[TMB]   TAKAYASU, M ; MINERVINI, J V. ; BROMBERG, L ; RUDZIAK, M K. ; WONG, T: Investigation of Twisted and Stacked Tape Cable Conductor. In: *CEC-ICMC 2011*. Spokane,

[TMB11]   TAKAYASU, M ; MINERVINI, J V. ; BROMBERG, L ; RUDZIAK, M K. ; WONG, T: Investigation of Twisted Stacked-Tape Cable Conductor. In: *CEC-ICMC 2011*. Spokane, 2011

[TMC03]   TAKAHATA, K ; MITO, T ; CHIKARAISHI, H ; IMAGAWA, S ; SATOW, T: Coupling losses in cable-in-conduit conductors for LHD poloidal coils. In: *Fusion Engineering and Design* 65 (2003), 1, S. 39–45

[TMP85]   THOME, R J. ; MINERVINI, J V. ; PILLSBURY, R D. ; BECKER, H D. ; MANN, W R.: Safety and Protection for Large Scale Superconducting Magnets - FY1985 Report / Idaho National Engineering Laboratory. Idaho Falls, 1985. – Forschungsbericht. – 1–40 S

[TNE06]   TOBITA, K ; NISHIO, S ; ENOEDA, M ; SATO, M ; ISONO, T ; SAKURAI, S ; NAKAMURA, H ; SATO, S ; SUZUKI, S ; ANDO, M: Design study of fusion DEMO plant at JAERI. In: *Fusion Engineering and Design* 81 (2006), Februar, 8-14, S. 1151–1158. – ISSN 09203796

[TNE09]   TOBITA, K ; NISHIO, S ; ENOEDA, M ; KAWASHIMA, H ; KURITA, G ; TANIGAWA, H ; NAKAMURA, H ; HONDA, M ; SAITO, A ; SATO, S ; HAYASHI, T ; ASAKURA, N ; SAKURAI, S ; NISHITANI, T ; OZEKI, T ; ANDO, M ; EZATO, K ; HAMAMATSU, K ; HIROSE, T ; HOSHINO, T ; IDE, S ; INOUE, T ; ISONO, T ; LIU, C ; KAKUDATE, S ; KAWAMURA, Y ; MORI, S ; NAKAMICHI, M ; NISHI, H ; NOZAWA, T ; OCHIAI, K ; OGIWARA, H ; OYAMA, N ; SAKAMOTO, K ; SAKAMOTO, Y ; SEKI, Y ; SHIBAMA, Y ; SHIMIZU, K ; SUZUKI, S ; TAKAHASHI,

K ; TSURU, D ; YAMANISHI, T ; YOSHIDA, T: Compact DEMO, SlimCS: design progress and issues. In: *Nuclear Fusion* 49 (2009), Juli, 7, S. 075029. – ISSN 0029–5515

[TNT09]   TOBITA, K ; NISHIO, S ; TANIGAWA, H ; ENOEDA, M ; ISONO, T ; NAKAMURA, H ; TSURU, D ; SUZUKI, S ; HAYASHI, T ; TSUCHIYA, K: Torus configuration and materials selection on a fusion DEMO reactor, SlimCS. In: *Journal of Nuclear Materials* 386-388 (2009), April, S. 888–892. – ISSN 00223115

[TTT95]   TEN HAKEN, B ; TEN KATE, H H J. ; TENBRINK, J: Compressive and tensile axial strain reduced critical currents in Bi-2212 conductors. In: *Applied Superconductivity, IEEE Transactions on* 5 (1995), 2, S. 1298–1301

[TVP10]   TERZIEVA, S ; VOJENČIAK, M ; PARDO, E ; GRILLI, F ; DRECHSLER, A ; KLING, A ; KUDYMOW, A ; GÖMÖRY, F ; GOLDACKER, W: Transport and magnetization ac losses of ROEBEL assembled coated conductor cables: measurements and calculations. In: *Superconductor Science and Technology* 23 (2010), Januar, 1, S. 014023. – ISSN 0953–2048

[UIM09]   UEDA, H ; ISHIYAMA, A ; MIYAHARA, N ; KASHIMA, N ; NAGAYA, S: Radioactivity of YBCO and Bi-2223 Tapes Under Low Energy Neutron Flux. In: *Applied Superconductivity, IEEE Transactions on* 19 (2009), 3, S. 2872–2876

[Uso09]   USOSKIN, A: YBCO Coated Conductor Development at Bruker HTS: Status and Outlook. In: *Braunschweiger Supraleitungsseminar*. Braunschweig, 2009

[Uso10]   USOSKIN, A. *2G Wires and Conductors*. 2010

[UWB13a]   UGLIETTI, D ; WESCHE, R ; BRUZZONE, P: Fabrication Trials of Round Strands Composed of Coated Conductor Tapes. In: *Applied Superconductivity Conference*. Portland : EPFL-CRPP, 2013. – ISSN 1051–8223

[UWB13b]   UGLIETTI, Davide ; WESCHE, Rainer ; BRUZZONE, Pierluigi: Fabrication Trials of Round Strands Composed of Coated Conductor Tapes. In: *IEEE Transactions on Applied Superconductivity* (2013). – ISSN 1051–8223

[Vel08]   VELIKHOV, E. 80 years of fusion. In: *AAAS Annual Meeting*, Kurchatov Institute, Moscow, 2008

[VGT11]   VOJENČIAK, M ; GRILLI, F ; TERZIEVA, S ; GOLDACKER, W ; KOVÁČOVÁ, M ; KLING, a: Effect of self-field on the current distribution in Roebel-assembled coated conductor cables. In: *Superconductor Science and Technology* 24 (2011), September, 9, S. 095002. – ISSN 0953–2048

[VKS07]   VINOD, K ; KUMAR, R G A. ; SYAMAPRASAD, U: Prospects for MgB 2 superconductors for magnet application. In: *Superconductor Science and Technology*
20 (2007), Januar, 1, S. R1–R13. – ISSN 0953–2048

[Wal74]   WALTERS, C R.: *Design of Multistrand Conductors for Superconducting Magnet
Windings*. 1974

[War77]   WARK, K ; CLARK, B J. (Hrsg.) ; MAISEL, J W. (Hrsg.): *Thermodynamics*. 3.
McGraw-Hill, 1977. – ISBN 0–07–068280–1

[WAT87]   WU, M K. ; ASHBURN, J R. ; TORNG, C J. ; HOR, P H. ; MENG, R L. ; GAO, L ;
HUANG, Z J. ; WANG, Y Q. ; CHU, C W.: Superconductivity at 93 K in a New
Mixed-Phase Y-Ba-Cu-O Compound System at Ambient Pressure. In: *Phys. Rev.
Lett.* 58 (1987), 9, S. 908–910

[WBBG11]   WEISS, K-P ; BARTH, C ; BAGRETS, N ; GOLDACKER, W: HTS High Current
Cable: Issue of I_c Measurement. In: *MEM11 Workshop*. Okinawa, 2011

[WDT86]   WALTERS, C R. ; DAVIDSON, I M. ; TUCK, G E.: Long sample high sensitivity
critical current measurements under strain. In: *Cryogenics* 26 (1986), 7, S.
406–412

[Wei07]   WEISS, K-P: Final Report of the FBI Test Facility / Forschungszentrum Karlsruhe
GmbH (FZK). Eggenstein-Leopoldshafen, 2007. – Forschungsbericht

[WF53]   WIEDEMAN, G ; FRANZ, R: The thermal conductivity of metals. In: *Ann. Phys.*
89 (1853), S. 497–531

[WFDR12]   WERFEL, F N. ; FLOEGEL-DELOR, U ; ROTHFELD, R ; RIEDEL, T ; GOEBEL,
B ; WIPPICH, D ; SCHIRRMEISTER, P: Superconductor bearings, flywheels and
transportation. In: *Superconductor Science and Technology* 25 (2012), Januar, 1,
S. 014007. – ISSN 0953–2048

[WGF92]   WELP, U ; GRIMSDITCH, M ; FLESHLER, S ; NESSLER, W ; DOWNEY, J ;
CRABTREE, G W. ; GUIMPEL, J: Effect of Uniaxial Stress on the Superconducting
Transition in $YBa_2Cu_3O_7$. In: *Phys. Rev. Lett.* 69 (1992), 14, S. 2130–2133

[Wika]   WIKIPEDIA. *Hastelloy. In: http://en.wikipedia.org/wiki/Hastelloy*

[Wikb]   WIKIPEDIA. *Manganin. In: http://en.wikipedia.org/wiki/Manganin*

[Wikc]   WIKIPEDIA. *Poisson's ratio. In: http://en.wikipedia.org/wiki/
Poisson's_ratio*

[Wik07]   WIKIPEDIA.   *Deuterium-tritium fusion. In: http://de.wikipedia. org/w/index.php?title=Datei:Deuterium-tritium_fusion. svg&filetimestamp=20091128202729*. 2007

[Wil83]   WILSON, M N. ; SCURLOCK, R G. (Hrsg.): *Superconducting Magnets*. Oxford : Oxford University Press, 1983. – ISBN 0–19–854805–2

[Wil99]   WILSON, M N.: Superconductivity and accelerators: the good companions. In: *IEEE Transactions on Applied Superconductivity* 9 (1999), Juni, 2, S. 111–121. – ISSN 10518223

[Wil08]   WILLIS, P B.: Survey of Radiation Effects on Materials. In: *OPFM Instrument Workshop*, 2008

[WJW11]   WEISS, K-P ; JUNG, A ; WESTENFELDER, S ; VOSTNER, A ; JEWELL, M: Impact of cold work on ductility of ITER TF jacket material. In: *22nd International Conference on Magnet Technology*. Marseille, 2011

[WNT06]   WAGANER, L ; NAJMABADI, F ; TILLACK, M ; WANG, X ; ELGUEBALY, L: Design approach of the ARIES-AT power core and vacuum vessel cost assessment. In: *Fusion Engineering and Design* 80 (2006), Januar, 1-4, S. 181–200. – ISSN 09203796

[WPSL12]   WEI, C ; PAN, W ; SUN, S ; LIU, H: Irradiation effects on a glycidylamine epoxy resin system for insulation in fusion reactor. In: *Journal of Nuclear Materials* 429 (2012), Oktober, 1-3, S. 113–117. – ISSN 00223115

[WTM10]   WEIJERS, H W. ; TROCIEWITZ, U P. ; MARKIEWICZ, W D. ; JIANG, J ; MYERS, D ; HELLSTROM, E E. ; XU, a ; JAROSZYNSKI, J ; NOYES, P ; VIOUCHKOV, Y ; LARBALESTIER, D C.: High Field Magnets With HTS Conductors. In: *IEEE Transactions on Applied Superconductivity* 20 (2010), Juni, 3, S. 576–582. – ISSN 1051–8223

[WW11]   WEISS, K-P ; WESTENFELDER, S: Mechanical Investigation of TF Jacket Material (AB No 45001322) / KIT - ITEP - CRyoMaK. 2011. – Forschungsbericht

[WWJ11]   WEISS, K-P ; WESTENFELDER, S ; JUNG, A ; BAGRETS, N ; FIETZ, W: Determination of mechanical and thermal properties of electrical insulation material at 4.2 K. In: *AIP Conference Proceedings* 58 (2011), S. 148–155. ISBN 9780735410220

[Yam04]   YAMAMOTO, A: Advances in Superconducting Magnets for. In: *Applied Superconductivity, IEEE Transactions on* 14 (2004), 2, S. 477–484

[YLJ12]   YEN, F ; LIN, Q ; JIANG, D ; CHEN, X ; XU, Y ; WANG, J S. ; WANG, S Y.: Hydrostatic Pressure Effects on the Critical Current of YBCO Coated Conductor Wire. In: *IEEE Transactions on Applied Superconductivity* 22 (2012), 4, S. 8401103

[YMS81]   YAMAMOTO, S ; MAENO, M ; SUZUKI, N ; AZUMI, M ; TOKUDA, S ; KATAGIRI, M ; SENGOKU, S ; YAMAUCHI, T ; KUMAGAI, K ; TAKEUCHI, H ; SUGIE, T ; KAWAKAMI, T ; OHASA, K ; MATSUDA, T ; KIMURA, H ; MATSUMOTO, H ; ODAJIMA, K ; HOSHINO, K ; YAMAMOTO, T ; KONOSHIMA, S ; KURITA, G ; AMANO, T ; OKAMOTO, M ; SHIMIZU, K ; SHIMOMURA, Y: Magnetohydrodynamic activity in the JFT-2 tokamak with high-power neutral-beam-injection heating. In: *Nuclear fusion* 21 (1981), 8, S. 993

[YOIS11]   YAMADA, Y ; OHKUMA, T ; IZUMI, T ; SHIOHARA, Y: Bending strain characteristics of RE123 Superconductors. In: *MEM11 Workshop*. Okinawa, 2011

[YSL05]   YI, H P. ; SONG, X H. ; L, Liu ; R, Liu ; ZONG, J ; ZHANG, J S. ; ALAMGIR, A K M. ; LIU, Q ; HAN, Z ; ZHENG, Y K.: Development of HTS BSCCO wire for power applications. In: *IEEE Transactions on Applied Superconductivity* 15 (2005), 2, S. 2507–2509

[YTK10]   YOSHIDA, K ; TSUCHIYA, K ; KIZU, K ; MURAKAMI, H ; KAMIYA, K ; PEYROT, M ; BARABASCHI, P: Design and Construction of JT-60SA Superconducting Magnet System. In: *J. Plasma Fusion Res.* 9 (2010), S. 214–219

[ZGRS07]   ZHOU, Y X. ; GHALSASI, S ; RUSAKOVA, I ; SALAMA, K: Flux pinning in MOD YBCO films by chemical doping. In: *Superconductor Science and Technology* 20 (2007), September, 9, S. S147–S154. – ISSN 0953–2048

[ZIB92]   ZHUKOVSKY, A ; IWASA, Y ; BOBROV, E S. ; LUDLAM, J ; WILLIAMS, J E C. ; HIROSE, R ; ZHAO, Z P.: 750 MHz NMR magnet development. In: *Magnetics, IEEE . . .* 28 (1992), 1, S. 644–647

[ZLE05]   ZEISBERGER, M ; LATKA, I ; ECKE, W ; HABISREUTHER, T ; LITZKENDORF, D ; GAWALEK, W: Measurement of the thermal expansion of melt-textured YBCO using optical fibre grating sensors. In: *Superconductor Science and Technology* 18 (2005), Februar, 2, S. S202–S205. – ISSN 0953–2048

[Zoh10]   ZOHM, H: On the minimum size of DEMO. In: *Fus. Sci. Technology* (2010), S. 613–624

[Zoh11]   ZOHM, H: Update on Topic P1: Steady State Operation. In: *3rd Meeting of the German DEMO working group*. Jülich, 2011

Karlsruher Schriftenreihe zur Supraleitung
(ISSN 1869-1765)

Herausgeber: Prof. Dr.-Ing. M. Noe, Prof. Dr. rer. nat. M. Siegel

**Die Bände sind unter www.ksp.kit.edu als PDF frei verfügbar oder
als Druckausgabe bestellbar.**

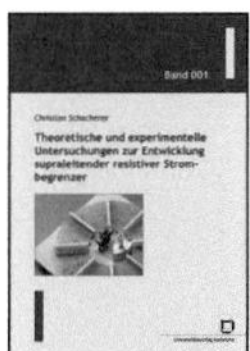

Band 001
Christian Schacherer
**Theoretische und experimentelle Untersuchungen zur
Entwicklung supraleitender resistiver Strombegrenzer.** 2009
ISBN 978-3-86644-412-6

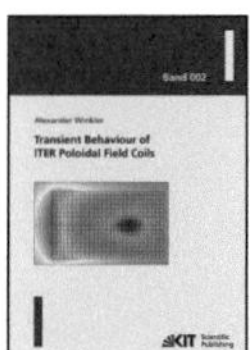

Band 002
Alexander Winkler
Transient behaviour of ITER poloidal field coils. 2011
ISBN 978-3-86644-595-6

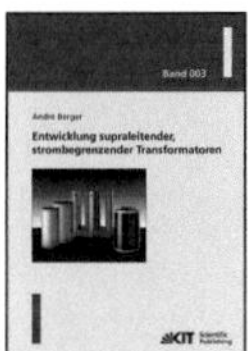

Band 003
André Berger
**Entwicklung supraleitender, strombegrenzender
Transformatoren.** 2011
ISBN 978-3-86644-637-3

Band 004
Christoph Kaiser
**High quality Nb/Al-AlOx/Nb Josephson junctions.
Technological development and macroscopic quantum
experiments.** 2011
ISBN 978-3-86644-651-9

Band 005
Gerd Hammer
**Untersuchung der Eigenschaften von planaren Mikrowellen-
resonatoren für Kinetic-Inductance Detektoren bei 4,2 K.** 2011
ISBN 978-3-86644-715-8

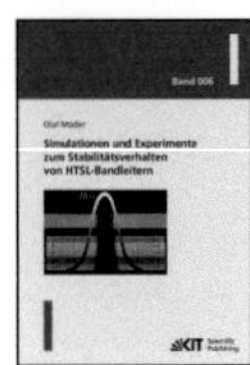

Band 006
Olaf Mäder
**Simulationen und Experimente zum Stabilitätsverhalten
von HTSL-Bandleitern.** 2012
ISBN 978-3-86644-868-1

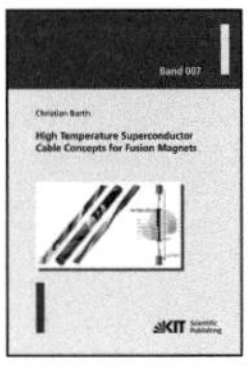

Band 007
Christian Barth
**High Temperature Superconductor Cable Concepts for
Fusion Magnets.** 2013
ISBN 978-3-7315-0065-0